Life Without a Manager

Autopoiesis from Cells to Civilizations

NBM House

Life Without a Manager: Autopoiesis from Cells to Civilizations

NBM House

Published by NBM House, 2026.

LIFE WITHOUT A MANAGER: AUTOPOIESIS FROM CELLS TO CIVILIZATIONS

First edition. March 18, 2026.

ISBN: 979-8232010379

Written by NBM House.

Life Without a Manager: Autopoiesis from Cells to Civilizations

Introduction: The Strange Claim

Start with a death. Not a dramatic one—no organism convulsing, no visible wound. Just a cell, a bacterium, that has stopped maintaining its own boundary. Its chemistry continues for a moment; the molecules are all still there. Nothing has been subtracted. And yet the cell is dead, and it will never be alive again, and the difference between the living state and the dead one turns out to be almost impossible to locate in physical terms.

This is where the trouble begins.

For centuries, biology accumulated lists. Life, it was said, involves metabolism. Reproduction. Response to stimuli. Homeostasis. Growth. These lists are not wrong—they describe real patterns—but they share a problem. Each item can be found in something we would not call alive. Fire metabolizes. Crystals grow. Thermostats respond. Prions reproduce, in a manner of speaking. Add enough qualifications and the list either becomes too narrow to capture some genuine edge cases, or too broad to exclude the obviously nonliving. The criteria keep failing at the margins, and the margins are exactly where the interesting questions live.

The deeper problem is that lists describe features. What biology has struggled to identify is the underlying process that produces and maintains those features. The standard move in twentieth-century biology was to go smaller: find the mechanism, reduce to chemistry, explain the cell in terms of its molecular components. This worked extraordinarily well as a research program. It produced the double helix, the genetic code, a detailed understanding of metabolic pathways that still generates Nobel prizes. What it did not produce was an answer to the original question. Knowing every protein in a living cell does not tell you what it means for the cell to be alive, any more than knowing every word in a sentence tells you what the sentence means.

In the late 1960s and early 1970s, two Chilean biologists—Humberto Maturana and Francisco Varela—decided to take the problem seriously in a different way. Their background was in neurophysiology and cognitive science, not evolutionary theory or molecular biology, and this gave them an unusual angle of approach. They were not asking how life evolved or what chemicals it required. They were asking what kind of thing a living system is. What is the organizational logic that makes it live, rather than the material inventory of its parts?

The concept they arrived at is called autopoiesis. The word is constructed from Greek—autos, self; poiesis, creation or production—and it names something precise: a system that produces the components of its own organization through its own processes, using those same components. A living cell does not receive its structure from outside. It continuously generates itself. The membrane is produced by processes occurring inside the membrane. Those processes are sustained by the membrane's capacity to regulate what passes through. There is no external agent directing this. There is no blueprint being read. There is only the loop, maintaining itself against the disorder that would otherwise claim it.

The loop is the thing. This is what autopoiesis is: not a substance, not a property, not a blueprint, but a process that continuously recreates the conditions for its own continuation. When that loop breaks, the organism dies—not because something has been removed, but because the self-generating organization has collapsed. The matter remains. The life does not.

This may sound like a narrow technical point, a refinement in biological theory. It is not. Once you accept the autopoietic definition of life, its consequences start moving outward in ways that are difficult to contain. If life is an organizational loop rather than a special substance, then cognition—perception, sense-making, response to the world—is not something that requires brains. It is, in some form, present wherever autopoiesis is. Every bacterium navigating a chemical gradient is making distinctions that matter for its survival. It is not thinking in any rich sense, but it is doing something that occupies the same functional space: it is differentiating its world according to what sustains the loop and what threatens it. Call that cognition if you like, or call it something else, but notice that it falls out naturally from the definition of life and requires no additional explanation.

From there, the ideas keep propagating. If identity—the boundary between self and world—is something that must be continuously maintained rather than simply possessed, then the self is not a thing but an activity. This applies to cells and, the argument will suggest, to minds. And if systems can be autopoietic without being biological—if the logic of self-production can operate in social institutions, in legal structures, in forms of collective organization—then the concept has left biology entirely and become something closer to a general theory of what it means for a complex system to persist.

Niklas Luhmann, the German sociologist who spent thirty years applying autopoietic theory to social systems, believed that society could be understood as a self-producing communication loop: it generates the elements of which it is composed, and those elements reproduce the system. People are not components of this system; they are its environment. The institution does not exist to serve its members. It exists to continue existing. This sounds cynical, and in some applications it is, but it is primarily descriptive—a formal account of why organizations outlive their founders, why legal systems resist reform, why institutions behave in ways that frustrate the individuals inside them. The system is not malicious. It is autopoietic.

None of this is easy to sit with. The autopoietic framework is in certain respects a set of dissolutions. It dissolves the intuition that life is a special substance. It dissolves the boundary between perception and action—between an organism registering the world and the organism constituting the world it registers. It dissolves the stable, unified self into a process that can succeed or fail. It puts the observer inside the system being observed, which creates problems for objectivity that cannot simply be noted and set aside. Maturana, in particular, drew radical conclusions from this observer-dependence: that there is no objective reality accessible from outside some perspective, only operational viability within one. Whether you follow him that far is a choice this book will not make for you. But you will need to understand what is at stake.

What is at stake, ultimately, is the question of what it means for anything to sustain itself in time. Cells do it. Minds may do it. Institutions do it. And as artificial systems become more capable of modifying their own architecture and maintaining their own processes, the question will arise—possibly uncomfortably—of whether the loop has closed in a domain we thought we controlled.

The cell has been doing this for roughly four billion years. It does not know it is remarkable. We are just beginning to understand why it is.

Chapter 1: Two Biologists Against the Machine

In 1943, Erwin Schrödinger gave a series of lectures in Dublin that would go on to influence a generation of biologists. The lectures were published the following year as a short book called What is Life?, and their central claim was that the living cell must contain something like an aperiodic crystal—a molecular structure capable of storing the information required to specify an organism. Schrödinger was a physicist, not a biologist, and the book had the usual physicist's virtues: it was clear, it was bold, and it was productively wrong in ways that generated real research. Francis Crick later credited it as a significant influence on his path toward DNA.

What is Life? established a frame that dominated biology for decades. Life, on this view, was fundamentally an information problem. The cell was a system for reading and executing a program. The interesting questions were about the code: how it was stored, how it was copied, how it was translated into proteins. These questions turned out to be extraordinarily productive. By the 1960s, molecular biology had cracked the genetic code, mapped the central dogma—DNA to RNA to protein—and begun the long project of reading genomes that continues today. The reductionist program was not just successful; it was one of the most successful research programs in the history of science.

But it left something out, and the thing it left out was the question Schrödinger had originally asked. The title of his book was not What Does a Cell Contain? or How Is Genetic Information Structured? It was What is Life? The reductionist program answered the mechanistic questions in detail and quietly set the original question aside. By the 1960s, asking what it meant for a cell to be alive—as opposed to asking how its machinery worked—was beginning to seem almost philosophically naive, the kind of question that would dissolve once the mechanisms were understood.

Humberto Maturana was not persuaded. He was a Chilean neurobiologist working in the late 1960s on the neurophysiology of vision, specifically on how the frog's visual system responded to color. The standard account assumed the visual system was a device for representing external reality: light hits the retina, signals travel through the optic nerve, the brain constructs an internal image of the world. Maturana's experiments suggested something more complicated. The frog's visual system did not seem to be building accurate representations of a color-defined world outside. It was producing internal states that bore a systematic but indirect relationship to external conditions. The frog was not a camera. It was doing something else, something harder to name.

This bothered Maturana for reasons that went beyond neuroscience. If the nervous system was not a representation-building device, what was it? What were the criteria for its operation? What was it organized to do? He found he could not answer these questions without first answering a prior one: what is the organization of any living system, considered as such? What makes it the kind of thing it is?

Francisco Varela came to the problem from a different direction. He was Maturana's student, a mathematician and biologist with a philosophical temperament and a serious interest in the foundations of logic—he would later work on Spencer-Brown's Laws of Form, a text concerned with the formal properties of self-referential systems and the act of drawing distinctions. Varela was drawn to the mathematics of circularity: systems that referred to themselves, processes that produced their own conditions of operation. He saw in Maturana's neurophysiological puzzlement something that called for a formal solution, not just a biological one.

Together, in the late 1960s and early 1970s, they developed what would become autopoietic theory. The collaboration was not always smooth—their intellectual styles differed, and they would later diverge on several philosophical conclusions—but in this period they were asking the same question with enough shared vocabulary to make progress. The core question was: what is the organization that is both necessary and sufficient for a system to be living?

To understand why this question was difficult, it helps to understand the landscape they were working against.

Cybernetics, which had emerged in the 1940s from Norbert Wiener's work on feedback control, offered one influential framework. A cybernetic system was defined by its goal-directed behavior: it used negative feedback to reduce the difference between its current state and a target state. The thermostat was the canonical example—it measured temperature, compared it to a set point, and activated a heater or cooler to close the gap. More complex cybernetic systems could pursue more complex goals, adjust their behavior in response to outcomes, learn in a limited sense. The framework was powerful enough to unify work in engineering, biology, and early computer science under a common vocabulary.

But cybernetics took the goal as given from outside. Someone set the thermostat. Someone specified what the system was supposed to do. The loop existed, but it had been designed by an external agent. For living systems, there was no external designer. The cell's organization was not given to it; it was the cell's own. Cybernetics could describe a living system's behavior after the fact, but it could not account for the origin of that behavior's organizational logic. It presupposed exactly what needed to be explained.

General systems theory, associated with Ludwig von Bertalanffy, offered a broader framework: the idea that certain organizational principles—feedback, hierarchical structure, open exchange with the environment—applied across biological, physical, and social systems. This was more general than cybernetics and less tied to engineering metaphors, but it faced a related problem. It described properties that living systems had without identifying what made them living rather than merely complex. An open system exchanging matter and energy with its environment was a useful description; it just applied equally well to a whirlpool and a bacterium.

Information theory, emerging from Claude Shannon's 1948 paper, gave biology a new vocabulary in the 1950s and 1960s. Genetic information, coded messages, noise and signal—the metaphors were productive and in some respects precise. But Shannon's information was defined entirely in terms of probabilities and symbol sequences, without reference to meaning or function. A gene had information content in Shannon's sense regardless of what protein it encoded or whether that protein did anything useful. The theory described the transmission and storage of patterns but said nothing about why some patterns mattered and others did not. For biology, meaning was the entire point.

Maturana and Varela's diagnosis was that all of these frameworks—cybernetics, systems theory, information theory—shared a common assumption: they modeled living systems as input-output devices. Something came in from the environment, the system processed it, something went out. The environment specified the input; the system's internal organization determined the output. This seemed natural, because it was how most designed systems worked. A radio receives a signal and produces sound. A computer receives instructions and produces results. The model was intuitive and mathematically tractable.

The problem was that living systems did not, on close examination, work this way. A cell does not receive instructions from outside and execute them. Its genome is not a program running on a processor; it is a component of an ongoing chemical process, and what the genome does depends on the state of the cell, which depends on what the genome has already done, which depends on the cell's history, and so on in a loop that has no clean entry point. The environment does not instruct the cell; it perturbs it. The cell responds not according to the environmental signal's content but according to its own organization. Two cells with different internal states can receive identical chemical signals and respond entirely differently. The signal does not determine the response; the cell's structure does.

This was the observation that forced Maturana and Varela toward a new framework. If the cell was not an input-output device, what was it? Their answer was: it was a self-producing system. It did not process external instructions. It maintained itself. The relevant loop was not between the system and its environment but inside the system itself: the processes that produced the system's components were the same processes sustained by those components. The organization was circular all the way down, and that circularity was not a defect or a paradox to be resolved—it was the definition of life.

Maturana first articulated the concept in a 1970 paper on the neurophysiology of cognition, in Spanish, which circulated among a small group of researchers. He and Varela developed it more formally in a 1972 monograph called Autopoiesis and Cognition, published initially in Chile and later, in 1980, by Reidel in the Netherlands as part of a series in the philosophy of biology. The reception was mixed. Biologists tended to find it too philosophical; philosophers tended to find it too biological; and the scientists in adjacent fields who might have bridged the gap were mostly committed to the input-output model the framework was questioning. The book sold modestly.

What it said, though, was genuinely novel. Not in the way that scientific novelty usually works—a new measurement, a new mechanism, a new structure identified in nature—but in the way that conceptual novelty works: it relocated where you were supposed to look. Instead of asking what a cell was made of, or what information it contained, or what goals it was pursuing, Maturana and Varela asked what kind of organization produced the phenomenon of life. The answer they gave has not been superseded, though it has been extended, contested, and applied in directions neither of them anticipated.

The bacterium in a petri dish, doing what it has always done, is not reading a program. It is not executing instructions. It is maintaining a loop. That loop is what makes it alive. Everything else—the DNA, the proteins, the membrane channels, the metabolic pathways—is in service of that. The loop is the thing that the reductionist program, for all its power, could not quite find, because it was looking at the components and not at the organization the components produced.

Maturana and Varela were looking at the organization. That turned out to matter.

Chapter 2: What Autopoiesis Actually Is

Here is the formal definition, from Maturana and Varela's 1972 monograph: an autopoietic system is organized as a network of processes of production of components that produces the components which, through their interactions and transformations, continuously regenerate and realize the network of processes that produced them; and constitute it as a concrete unity in space.

This sentence is worth reading twice. It is dense by necessity, not by affectation. Each phrase is doing specific work, and unpacking them in sequence is the fastest route to understanding what the concept actually claims.

Start with "network of processes." Not components. Not a structure. A network of processes. The emphasis is on what the system does, not what it is made of. A car engine and a living cell may share some of the same atomic constituents, but what distinguishes the cell is not its material inventory—it is the organized set of ongoing transformations occurring within it. The system, at any given moment, is the current realization of a process that extends backward through time and is reaching forward, attempting to continue.

Now add "production of components." The processes in the network are productive—they make things. Specifically, they make the components of the network itself. Enzymes catalyze reactions that produce other enzymes. Lipid molecules assemble into membranes that contain the reactions producing more lipid molecules. This is the first hint of the loop: the system is making its own parts.

"Continuously regenerate and realize the network." The components that are produced do not simply accumulate. They feed back into the network of processes, sustaining it. The products are the producers. This is the loop completed: processes produce components; components sustain processes; processes produce components. There is no exit from this circle, and there is no entry point from outside. The system is, in a precise sense, self-producing.

Finally: "constitute it as a concrete unity in space." This is the boundary condition, and it is not incidental. An autopoietic system must produce and maintain a physical boundary that distinguishes it from its environment. Without that boundary, the network of processes would simply dissolve into the surrounding medium. The boundary is not imposed from outside; it is produced by the same network it encloses. In a cell, the membrane is synthesized by the internal chemistry, and it is the membrane that makes the internal chemistry possible by keeping the relevant molecules concentrated and separated from the exterior. The boundary creates the inside; the inside creates the boundary.

Three features follow directly from this definition, and each one matters.

The first is operational closure. An autopoietic system is closed, organizationally, in the sense that all its processes refer back to the system itself. This is not the same as being physically closed—cells exchange matter and energy with their environment constantly. It means that the system's operations are defined with reference to the system's own organization, not with reference to external specifications. The cell does not receive instructions about how to be a cell. Its operations are determined by its current state, which is the product of its history of self-production. Nothing from outside can directly specify what the cell does; it can only perturb a system that will respond according to its own internal logic.

The second feature is that identity and organization are the same thing. An autopoietic system has an identity—it is this system, distinguishable from everything else—and that identity just is its ongoing self-production. There is no self that exists independently of the process. The system is the loop. When the loop stops, the identity stops. This is why cell death is so philosophically clean: the matter persists, the identity does not, and the difference is precisely the cessation of autopoietic organization.

The third feature is that structure and organization must be distinguished. The organization of an autopoietic system is the set of relations among processes that makes it autopoietic—the abstract pattern that defines what kind of system it is. The structure is the particular physical realization of that organization at any given moment—these specific molecules, this specific configuration, here and now. The same organization can be realized in different structures; two cells can have the same autopoietic organization while differing substantially in their molecular composition. And the structure is always changing, because the processes are always running, replacing components, adjusting concentrations. What stays constant, as long as the system is alive, is the organization. What changes continuously is the structure that realizes it.

This distinction resolves an apparent paradox. A living cell replaces most of its molecular components over time—the phospholipids in its membrane, the proteins in its cytoplasm, much of its nucleic acid machinery. In a meaningful material sense, it is not the same cell it was a week ago. Yet it is, in an equally meaningful sense, the same cell: the same organism, with the same identity. Autopoietic theory says this is not a paradox at all. What persists is not the material but the organization. The cell is continuous with its past self in exactly the way that matters: the self-producing loop has been maintained without interruption. Everything else is logistics.

Now consider two systems that are often mistaken for autopoietic, and examine why they fail the test.

A factory produces components. Some factories even produce components that go into making other factories. This seems superficially similar to autopoiesis—a productive system, generating its own kind. But a factory does not produce the components of its own ongoing operation from within itself. The raw materials come from outside. The energy comes from outside. The specifications for what to produce come from outside. The workers who maintain the machinery exist independently of the factory's productive processes. Most critically, the factory does not produce the factory. Its products leave and go elsewhere. The productive loop does not close back on the system producing it. A factory is what Maturana and Varela call an allopoietic system: it produces something other than itself.

A flame is a more interesting case, and more instructive. A flame does maintain itself through a process of ongoing chemical transformation. It consumes fuel, produces heat and combustion products, and the heat sustains the reactions that consume more fuel. There is something loop-like here. A flame can, in a limited sense, reproduce—touch a candle to another candle and you have two flames. And a flame has something like a boundary: a surface region where the reaction front sits.

But a flame does not produce its own components. The fuel is external. The oxygen is external. The flame does not synthesize the molecules it consumes; it destroys them. More precisely, a flame has no components in the relevant sense—it is a process without a network of component-producing processes underlying it. It is a sustained chemical reaction, not a self-producing organization. Deprive it of either fuel or oxygen and it stops immediately, with nothing left that might resume given different conditions. A cell deprived of nutrients can enter dormancy, slow its metabolism to near-zero, and persist for extended periods in a state that is genuinely borderline but genuinely alive. The flame has no such resources. When the reaction stops, there is nothing there to continue.

The flame fails on the boundary condition too. The reaction front is a consequence of physical dynamics, not a produced boundary. It does not constitute an inside and an outside. There is no interior of a flame that the flame is maintaining.

These comparisons clarify the specificity of the autopoietic claim. It is not enough to be self-sustaining in some loose sense. It is not enough to involve feedback. It is not enough to be a process rather than a static structure. The claim is precise: *the components that sustain the process must be produced by the process they sustain, within a boundary that is itself produced by that process.* All three conditions—component production, organizational closure, and self-produced boundary—must be satisfied simultaneously. The flame satisfies none of them cleanly. The factory satisfies none of them at all.

One further distinction is worth making, because it becomes important in later chapters. Maturana and Varela were careful to say that autopoiesis defined the organization of living systems, not their full complexity. A cell can be autopoietic without doing anything particularly interesting by the standards of neuroscience or evolutionary biology. Autopoiesis is the floor, not the ceiling. It is what makes a system count as alive. What a living system does beyond maintaining its autopoiesis—how it behaves, how it evolves, how it interacts with other systems—is a separate set of questions that autopoietic organization makes possible but does not determine.

This matters because the concept is sometimes criticized for being too minimal—for reducing life to a bare loop that seems to leave out most of what makes life interesting. The criticism misunderstands the project. Maturana and Varela were not trying to explain everything about living systems. They were trying to identify the organizational invariant—the thing that is common to all living systems and necessary to all of them—and to give it a rigorous characterization. That the characterization is austere is not a defect. It is what makes it precise.

The production loop, maintained within a self-generated boundary, with all processes referring back to the network they compose: this is what any living thing is doing, right now, as you read this. The bacterium does it with three thousand proteins and a lipid bilayer two molecules thick. The human cell does it with twenty thousand protein-coding genes, elaborate organelles, and a membrane studded with thousands of receptor types. The complexity differs by orders of magnitude. The organizational logic is the same.

What is striking, when you stay with the definition long enough, is how much it removes rather than adds. There is no program being run. There is no goal being pursued in any externally specified sense. There is no information being received from outside and processed into outputs. There is just the loop—sustaining itself, replacing its own parts, maintaining its own boundary—and doing this so reliably, across such enormous ranges of environmental variation, that we have been watching it for four billion years without it stopping.

That is a long time to maintain a loop. It suggests the loop is doing something right.

Chapter 3: The Cell as a Philosophical Weapon

There is a tendency, when thinking about life, to start from the top. The organism that presents itself most vividly to our attention is the organism most like us: complex, motile, possessed of something resembling a nervous system, operating in an environment we recognize as our own. We reason about life by analogy to ourselves, and when that gets too anthropomorphic we retreat one step to mammals, then to vertebrates, then to multicellular animals in general. The single cell tends to appear late in this sequence, as a simplified case, a stripped-down version of the real thing.

Autopoietic theory inverts this. The single cell is not a simplified organism. It is the purest instance of the phenomenon that needs explaining. Everything that is essential to life is present in a bacterium with no nervous system, no organs, no developmental program, no evolutionary history of multicellularity. If you want to understand what life organizationally is, the bacterium is where to look, because it has nothing else going on. It is the loop with no additional structure obscuring the loop.

A prokaryotic cell—the kind that appeared on Earth roughly 3.5 billion years ago and still constitutes the majority of living biomass—is approximately one to ten micrometers in diameter. It has no nucleus. Its single circular chromosome floats free in the cytoplasm. It has no mitochondria, no endoplasmic reticulum, none of the elaborate internal membrane architecture that eukaryotic cells possess. What it has is: a membrane, a genome, ribosomes to translate that genome into proteins, a set of metabolic enzymes to extract energy and synthesize molecules, and the ongoing process that keeps all of these in relation to each other. That is enough. It has been enough for four billion years.

The membrane deserves particular attention, because it is where the autopoietic logic becomes most tangible. A bacterial cell membrane is a phospholipid bilayer: two sheets of lipid molecules arranged tail-to-tail, with their hydrophilic heads facing outward into the aqueous environment and inward toward the cytoplasm. This structure is thermodynamically favorable—the molecules spontaneously arrange this way in water—but the arrangement alone does not constitute a living boundary. What makes it a living boundary is that the cell produces the phospholipids that compose it. The membrane is not a container into which life has been placed. It is a product of the metabolic processes it encloses, continuously renewed as its components are degraded and replaced.

Embedded in this membrane are proteins: channels that regulate what passes in and out, receptors that respond to external chemical signals, pumps that maintain electrochemical gradients by moving ions against their concentration gradient at the cost of energy. Each of these proteins is produced by the cell's internal processes—transcribed from DNA, translated by ribosomes, inserted into the membrane. The membrane is thus not a fixed structure but a dynamically maintained one, produced and sustained by the chemistry it separates from the outside world.

This is the autopoietic loop made visible. The membrane keeps the metabolic chemistry concentrated and separated from the environment, making it possible. The metabolic chemistry produces the membrane, making it possible. Neither has priority. Neither is the cause and the other the effect. They are mutually constitutive, and the system that emerges from their mutual constitution is the cell.

Now consider metabolism more carefully. The word refers to the complete set of chemical transformations occurring within a living system—the breakdown of nutrients into usable energy, the synthesis of the molecules the cell needs, the disposal of waste products. In a bacterium, this network involves several hundred to several thousand distinct chemical reactions, each catalyzed by a specific enzyme, each enzyme produced by the cell's own genetic machinery. The network is not a pipeline with a start and an end. It is a web: the product of one reaction is the substrate of another, energy liberated in one pathway drives reactions in another, and feedback loops adjust the rates of individual reactions in response to the concentrations of their products.

What the metabolic network as a whole does, from an autopoietic perspective, is produce the molecules required to maintain the metabolic network. Enzymes catalyze reactions that produce the precursors of other enzymes. The energy extracted from nutrients powers the synthesis of the molecules that extract energy from nutrients. This is not a vicious circle. It is an operational one: a productive loop that sustains itself by doing what it does. The circularity is the point.

There is a concept from theoretical biology that captures this: *organizational closure.* A metabolic network is organizationally closed when every molecular species required for the network's operation is either produced by the network or constitutes a boundary condition for it. In a living cell, this closure is never absolute—cells require inputs of matter and energy from their environment—but the *organization* of the metabolic network is closed: the relations among processes are self-referential, defined in terms of the network itself rather than in terms of any external specification. The cell does not consult an external template to know what reactions to run. It runs the reactions it runs because it is the cell it is, and it is the cell it is because it runs those reactions.

What does it mean, then, for a cell to die?

The conventional answer is that death involves damage: to the genome, to the membrane, to critical proteins. And damage is often the proximate cause. But the autopoietic account says something more precise: death is the irreversible breakdown of the self-producing organization. The cell dies when the loop cannot maintain itself—when the processes that produce components fall below the threshold required to replace what is being degraded, or when the membrane's integrity fails and the internal chemistry disperses into the environment, or when a critical enzyme is lost and cannot be resynthesized.

Notice what this framing removes. Death is not the loss of a special substance—no vital force departs. It is not the corruption of an information-bearing medium, though that may be part of the causal story. It is the cessation of a process. What was alive is now a collection of molecules, still following the laws of chemistry, still capable of reacting with each other, but no longer organized into a self-producing loop. The molecules will continue to interact; the organization will not reconstitute itself. The matter remains; the life does not.

This has an implication that is worth pausing on. If life is a process rather than a substance or a property of matter, then there is a sense in which asking where life is located is the wrong question. Life is not in the DNA, though the DNA is necessary for it. It is not in the membrane, though the membrane is necessary for it. It is not in the enzymes, though without the enzymes it cannot proceed. It is in the relations among all of these—in the operational organization that makes each of them what they are within the system. Remove any one component and the organization may collapse. But the organization itself is not a component. It is the pattern that the components collectively realize.

This is what Maturana and Varela meant by calling the cell a unity. Not that it is simple, but that it has a mode of existence as a whole that is not reducible to its parts. The cell is not the sum of its molecules. It is the organization those molecules maintain. And that organization has a kind of causal priority: it is because the cell has the organization it has that the molecules within it behave as they do, rather than diffusing away or reacting randomly. The organization constrains the chemistry; the chemistry realizes the organization.

Multicellular organisms complicate this picture without overturning it. A human body contains roughly thirty-seven trillion cells, each of which is independently autopoietic—each maintaining its own membrane, its own metabolism, its own self-producing loop. The coordination of these cells into tissues, organs, and organ systems involves additional layers of organization: signaling molecules, developmental programs, immune surveillance, the whole machinery of multicellular life. But none of this replaces cellular autopoiesis. It builds on top of it. The liver cell doing its work in your body right now is autopoietic in exactly the same sense as the bacterium in a soil sample. It is maintaining its own loop while participating in larger organizational structures that depend on that maintenance continuing.

This nested structure—autopoietic components organized into higher-order systems—is something Varela would later call second-order autopoiesis, though the term is contested and the concept requires care. The key point is that higher-order organization does not absorb or dissolve the autopoiesis of the components. It depends on it. If the cells in your liver stop maintaining their autopoietic organization, the liver fails, and no amount of higher-order coordination can substitute for the basic loop.

There is something uncomfortable about taking the cell seriously as the fundamental unit of life, because it relocates where philosophical weight is supposed to sit. We are accustomed to thinking that the interesting questions about life are questions about organisms—about experience, agency, purpose, death as something that matters. The cell is background to all of this, the infrastructure, the thing we take for granted. Autopoietic theory does not say this is wrong. It says the background is where the essential operation is occurring, and that understanding organisms—understanding experience, agency, purpose—requires understanding what the background is actually doing.

The bacterium does not know it is remarkable. It is not trying to be a philosophical example. It is maintaining its loop, as it has been doing for roughly four billion years, indifferent to the theoretical apparatus we have been building to describe it. That indifference is itself instructive. The loop does not require our understanding. It does not require a nervous system to supervise it, a genome to consult for instructions, or an environment that cooperates. It requires only that the processes produce what the processes need, within a boundary the processes maintain.

Everything that follows in this book—cognition, identity, society, artificial minds, meaning—rests on taking this seriously. The cell is not a simplified version of something more interesting. It is the thing that all of the more interesting phenomena are versions of.

Chapter 4: Cognition Before Brains

Escherichia coli is a bacterium roughly two micrometers long, present in the gut of almost every mammal alive. It has no nervous system, no sense organs in any conventional meaning of the term, no neurons, no synapses, no brain. It has been studied more intensively than almost any other organism on Earth, and what researchers have found is that it navigates.

E. coli propels itself through fluid using rotating flagella—long helical filaments driven by molecular motors embedded in the cell membrane. When the motors rotate counterclockwise, the flagella bundle together and the cell moves in a roughly straight line, a behavior called a run. When one or more motors reverse to clockwise, the flagella fly apart, the cell tumbles randomly and reorients, and then the motors switch back and a new run begins in whatever direction the cell now faces. The result is a biased random walk: the cell cannot steer directly, but by extending runs when conditions are improving and shortening them when conditions are worsening, it moves up gradients of attractants and down gradients of repellents with genuine effectiveness.

The mechanism involves a network of signaling proteins that detect the concentration of specific chemicals in the environment and compare current concentrations to recent ones. The cell is not responding to absolute concentration—it is tracking the derivative, the rate of change. If aspartate concentration is rising, runs are extended. If it is falling, runs are shortened and tumbles come sooner. The comparison is temporal: the current state of the signaling network encodes, in a functional sense, a memory of the recent chemical past. The cell is computing something, in a minimal but precise way, and using the result to regulate behavior.

None of this requires invoking anything unusual about E. coli. It is unremarkable by bacterial standards. What is unusual is what it implies, once you look at it from an autopoietic perspective.

Maturana's claim, developed through the 1970s and stated most directly in Autopoiesis and Cognition, was that cognition is not a special capacity possessed by organisms with sufficiently complex nervous systems. It is a feature of all living systems. Specifically: living is cognizing. Any system that is autopoietic is, by virtue of being autopoietic, cognitive. The two are not identical—cognition is the broader claim about how the system relates to its environment—but they are inseparable. You cannot have one without the other.

This claim requires unpacking, because it depends on a definition of cognition that differs substantially from the ordinary one. In ordinary usage, cognition involves mental processes: perception, memory, reasoning, perhaps consciousness. Maturana's definition strips all of this away. Cognition, for him, is the activity of a living system in its domain of existence. It is what happens when an autopoietic system interacts with its environment in ways that are determined by the system's own organization rather than by the environment's properties. The cell that detects aspartate and adjusts its motility is not representing the aspartate or forming a concept of it. It is responding to a perturbation in a way structured by its own internal organization. That response counts as cognitive, in Maturana's sense, because it is an instance of a living system operating in its domain—making the distinctions it makes, doing what its organization does.

The key concept here is *sense-making*, a term developed more explicitly by Varela and later by the philosopher Evan Thompson and the biologist Terrence Deacon. A living system is not indifferent to its environment. It has a perspective on its environment constituted by what matters for its autopoietic maintenance. The chemical gradient that E. coli navigates is not a neutral physical fact about the world; it is a gradient of significance, structured by the cell's own needs. Aspartate is an attractant because it supports the cell's metabolism. Leucine is also an attractant, for related reasons. Certain metal ions are repellents because they interfere with cellular processes. These distinctions are not given by physics. They are given by the cell's organization.

This is what distinguishes sense-making from mere mechanical response. A thermostat responds to temperature, but the temperature is not significant for the thermostat in any meaningful sense. The thermostat has no stake in the outcome; it does not exist in a way that temperature fluctuations can threaten or sustain. The cell does. The distinction between attractant and repellent tracks, imperfectly but genuinely, the distinction between what supports autopoietic maintenance and what threatens it. The cell's perceptual world is organized around its own continued existence, because its perceptual world is constituted by the organization that makes it exist.

The contrast with representationalist accounts of cognition is sharp here, and it is worth being precise about why Maturana rejected the representationalist picture so firmly. On the standard cognitive science account that dominated in the 1970s and still commands significant allegiance today, cognition works by building internal representations of an external world and performing computations on those representations. The brain constructs a model of reality; behavior is generated by operations on that model. The environment provides inputs; the cognitive system processes them into outputs. This is the input-output framework applied to minds.

Maturana's neurophysiological work had already suggested this picture was wrong, at least as a complete account. His studies of the frog visual system showed that the retina was not transmitting a faithful image of the world to the brain. It was transmitting a highly processed, highly selective signal, structured by the retina's own organization—certain cells responding to moving edges, others to particular contrasts, others to specific combinations of features. The retina was doing something, not just conducting. And what it was doing was not, in the first instance, building a representation of the world. It was maintaining the activity of a neural network organized to sustain the organism's ongoing interaction with its environment.

The frog does not have a concept of fly and search for instances of it in the world. It has a visual system that responds to small moving objects in ways that are continuous with its history as an organism of its kind, and that response tends to produce behavior—snapping—that is survival-consistent. The frog is not wrong about the fly. But it is not representing the fly either. It is enacting a relation to the fly, through a nervous system whose organization is itself a product of the frog's autopoietic history.

Here the concept of structural coupling enters, one of the more technically important ideas in the Maturana-Varela framework. Two systems are structurally coupled when they have a history of interaction in which each has influenced the structural changes of the other, without either one determining those changes. The frog and the ecosystem it inhabits are structurally coupled: the frog's visual system has the organization it has partly because of the history of its lineage's interactions with environments containing insects and similar prey. The environment did not instruct the frog's visual system into being. But the history of structural interactions between frogs and their environments contributed to the structural changes—evolutionary changes, in this case—that produced it.

Structural coupling is also a present-tense process. The bacterium swimming up a gradient is structurally coupling with its chemical environment: each interaction perturbs the signaling network, and the history of perturbations shapes the network's current state. The individual cell has a chemical history, encoded in the current configuration of its signaling proteins, that reflects its recent environmental experience. This is not learning in any rich sense—the bacterium's memory is shallow and chemically simple—but it is a genuine instance of an autopoietic system being shaped by its history of coupling with an environment, in a way that influences its current behavior.

The claim that all living systems are cognitive met resistance from multiple directions, and the resistance was not unreasonable. Biologists objected that it diluted a useful technical concept to the point of uselessness—if cognition applies to bacteria, what does the word mean for humans? Cognitive scientists and philosophers of mind objected that it defined away the very features of cognition they found most interesting: intentionality, representation, the aboutness of mental states. What mattered, from their perspective, was not whether a system responded to its environment in organized ways, but whether it had mental states with content—states that were about things in the world.

These objections have some merit, and Maturana's response to them was not always fully satisfying. There is a genuine question about whether a concept of cognition broad enough to include bacteria is the right tool for understanding consciousness, language, or conceptual thought. The autopoietic answer is that you have to start somewhere, and the right place to start is with the most fundamental form of the phenomenon. A theory of cognition that applies only to vertebrates with large forebrains is not a theory of cognition—it is a theory of a very specific kind of nervous system activity. Understanding what is general about cognition requires understanding its minimal form, and the minimal form is present wherever there is life.

This is not a fully satisfying answer, but it is a principled one. And it has a productive consequence: once you accept that sense-making and self-production are internally related, the question of where cognition begins and physiology ends loses much of its grip. The bacterium navigating a gradient is not doing something categorically different from the neuron maintaining its membrane potential or the immune cell tracking an antigen. All of these are instances of an autopoietic system operating in its domain, making distinctions that matter for its maintenance, responding to perturbations in ways structured by its own organization. The complexity varies enormously. The organizational logic is continuous.

Varela, Thompson, and Rosch would develop these ideas further in their 1991 book *The Embodied Mind*, which brought the autopoietic account of cognition into dialogue with cognitive science, phenomenology, and Buddhist philosophy of mind. Their central term was *enaction*: cognition is not the processing of a pre-given world by an inner mechanism, but the enactment of a world and a mind through a history of structural coupling. The organism and its environment bring each other forth in their interaction. This is a stronger claim than Maturana's, and more philosophically contentious, but its roots are in the basic autopoietic picture.

For now, the key point is this: the claim that all living systems are cognitive is not a poetic inflation of bacterial behavior. It is the consequence of taking the autopoietic definition of life seriously and asking what follows from it. If a living system is defined by the production and maintenance of its own organization, and if that organization constitutes the domain in which the system operates, then the system's interactions with its environment are always structured by that organization—are always, in the functional sense, informed by what matters for the system's continuation. And that is cognition: not as a special gift of evolution, not as a property of sufficiently complex neural networks, but as the way any living system relates to the world it inhabits.

E. coli does not know it is navigating. It does not have experiences, in any sense we can responsibly attribute. But it is doing something that occupies the same logical space as what you are doing reading this sentence: orienting in an environment, making distinctions that matter, responding in ways structured by a history of being the kind of thing it is. The difference in complexity is real and enormous. The difference in kind is harder to locate than it first appears.

Chapter 5: Information Is Not What You Think

The word information has done more damage to the understanding of living systems than almost any other term borrowed from engineering. Not because it is wrong to use it—biologists and cognitive scientists use it every day with genuine precision—but because it carries assumptions that, once imported, are very difficult to dislodge. The assumptions are so deeply embedded in how we think about minds and cells that they feel like facts rather than choices.

The foundational assumption is this: information is a thing that exists in the environment, is detected by a living system, and is transmitted into the system where it causes effects. The environment sends; the organism receives. The message has content that the receiver decodes. This is the communication model—Shannon's model, more or less—and it is impeccably suited to its original purpose, which was characterizing the transmission of signals through noisy channels between a sender and a receiver who share a code. It is a model of engineering problems: how much can you compress a message, how much redundancy do you need to survive noise, what is the theoretical limit of a channel's capacity.

Transferred to biology, the model implies that the genome is a program, the cell is a computer running it, and the environment provides inputs that the program processes into behavioral outputs. Transferred to neuroscience, it implies that the sensory organs detect signals in the world, transduce them into neural codes, and send those codes upstream where the brain decodes them into perceptions. In both cases, the organism is a receiver and processor of information that originates outside it.

The autopoietic critique is not that this picture is entirely false. Cells do use chemical signals. Nervous systems do transduce and transmit. The critique is that the communication model misrepresents where the action is. It places the explanatory weight on the signal—on what the environment sends—when the explanatory weight belongs on the receiver's organization, which determines what effect, if any, any given perturbation will have.

Consider a concrete example. A cell is exposed to a molecule of cortisol. What happens? The cortisol binds to a receptor protein on the cell's surface or, because cortisol is lipid-soluble, passes through the membrane and binds to an intracellular receptor. The receptor-cortisol complex then interacts with the cell's gene regulatory machinery, altering which genes are transcribed and at what rate. The cell's behavior changes.

Now: did the cortisol carry information that the cell received and processed? In one sense, yes—cortisol reliably correlates with certain physiological states, and the cell's response to it is functionally appropriate to those states. But look more carefully at the causal structure. The cortisol molecule does not contain instructions. It does not specify what the cell should do. What it does is alter the conformation of a receptor protein, which alters that protein's interactions with other molecules, which propagates through the cell's regulatory network and produces effects. Those effects depend entirely on the cell's existing organization: what receptor proteins it expresses, how those proteins are connected to the regulatory network, what the current state of that network is, what genes are already being transcribed. A liver cell and an immune cell exposed to the same cortisol concentration will respond differently, not because they receive different information, but because they are differently organized.

This is the distinction Maturana insisted on, with characteristic terminological precision: the environment does not instruct the cell. It *perturbs* it. A perturbation is a change in the environment that triggers a state change in the system. What state change occurs is determined by the system's structure, not by the perturbation's content. The perturbation selects from among the state changes the system is capable of making; it does not specify which state change occurs. The cell is *structurally determined*: its behavior at any moment is a function of its current structure, which is the accumulated result of its autopoietic history. Nothing from outside can override this. The environment can trigger; it cannot program.

The same logic applies to nervous systems, though the complexity is vastly greater. When light strikes the retina, it does not transmit an image to the brain. It triggers a cascade of electrochemical changes in photoreceptor cells, which propagate through the retinal network—itself performing significant processing—and generate patterns of activity in the optic nerve. Those patterns are not pictures. They are not even straightforward representations of light distribution across the retina. They are the outputs of a network with its own complex organization, which has transformed the physical signal into something shaped by the retina's structure. The brain then receives these patterns and processes them through further networks whose organization is the product of developmental history, prior learning, current metabolic state, and attention. The final perceptual experience—seeing a red cup on a table—is produced by the visual system, not delivered to it.

None of this means the red cup isn't there. The autopoietic account is not idealism. The environment is real and it exerts genuine causal influence. But the nature of that influence is different from what the communication model implies. The environment does not send content that the organism unpacks. The organism's organization constitutes the space of possible responses to environmental perturbations, and the environment, in interacting with the organism, selects among them. The content—the meaning, the significance—is contributed by the organism. The world provides occasions; the organism provides the sense made of them.

This reframing has consequences for how we understand learning and development. The conventional picture of learning is that the environment provides information that the organism stores and uses to update its model of the world. The autopoietic alternative is that learning is a history of structural coupling: interactions with the environment that trigger structural changes in the organism, changes that alter how the organism will respond to future perturbations. The change is in the organism's structure, not in the content of an internal database. And the structural changes that occur are those the organism's current structure permits—you cannot learn what your current organization has no capacity to accommodate.

This is why the same experience can have entirely different effects on different people, and why the same person at different life stages is differently affected by encounters with the same environment. It is not that they receive different information. It is that their structural histories are different, and so the same perturbation triggers different structural changes. Education, trauma, practice, habituation—all of these are names for different kinds of structural coupling histories that alter the organism's organization in ways that persist and shape future responses.

Maturana introduced a term for the domain of interactions an organism can have without losing its autopoietic organization: the *consensual domain*. When two or more autopoietic systems interact recurrently, they can develop coordinated behaviors—not by transmitting representations to each other, but by mutually triggering structural changes that bring their behaviors into coordination. Language, on this account, is a consensual domain: a history of recurrent interaction between organisms in which certain behaviors have come to reliably trigger certain responses, not because those behaviors carry meaning in themselves, but because the organisms have been structurally coupled long enough that meaning has been built into their shared behavioral history. Words do not contain meanings. They trigger states in nervous systems that have been organized, through a history of coupling, to respond to them in particular ways.

This account of language is deliberately deflationary, and it provoked objections from linguists and philosophers of language who felt it left out everything interesting about meaning, reference, and truth. The objections are not without force. Maturana's account handles coordination well and handles content poorly. How a word can be about something in the world—how language can refer—is not obviously captured by a story about mutual structural coupling, and Maturana was not always careful about acknowledging this gap.

But the gap does not undermine the core point. The core point is that the communication model, applied naively to biological systems, inverts the actual causal structure. It treats the organism as a passive receiver of environmental content when the organism is an active, structurally determined system whose organization constitutes the space of possible responses to any perturbation. The environment matters. It matters enormously. But it matters by triggering, not by instructing. And what it triggers is always a function of what the system is, which is always a function of what the system has been.

There is an observer problem lurking here that will require its own chapter, but it is worth flagging now. If the organism's responses are structurally determined, and if the organism's structure is the product of its history, and if meaning is contributed by the organism rather than carried by the signal—then what is the status of the claim that the cortisol molecule carries information, or that the genome encodes the organism? These are descriptions produced by observers with their own organizational histories, using concepts developed within scientific communities that have their own structural coupling histories. The descriptions are not false, but they are not neutral either. They carry the observer's organization into the account.

Maturana took this seriously enough to develop an explicit epistemology around it, which will be addressed in Part III. For now it is enough to note the problem's presence. A theory of cognition that locates meaning-generation in the organism rather than in the environment eventually has to say something about the cognition involved in producing the theory itself. Autopoietic theory does not escape its own reach.

What it offers, before that complication arrives, is a cleaner account of how living systems actually relate to their environments. Not as receivers processing incoming content, but as structurally determined organizations for which the environment provides occasions to do what they do. The bacterium does not read the gradient. The frog does not decode the fly. The human nervous system does not reconstruct the scene. Each of these is an autopoietic system interacting with perturbations in ways that its organization—its history, its current structure, its maintained loop—makes possible and constrains.

The signal does not explain the response. The system does. This is the inversion autopoiesis performs on information theory, and it is a genuine one: not a rejection of the concept of information but a relocation of where the work gets done.

Chapter 6: Enacted Worlds

In 1909, the Baltic German biologist Jakob von Uexküll proposed an idea that most of his contemporaries found either obvious or scandalous, depending on their temperament. The idea was that each animal species inhabits a perceptual world of its own construction—not the physical world as such, but a world constituted by the sensory capacities and motor repertoires of that species. He called this world the Umwelt, from the German for environment or surrounding world, and he meant it precisely: the Umwelt is not the environment an organism happens to live in. It is the environment as the organism's own organization constitutes it.

A tick, Uexküll's canonical example, has three sensory modalities: sensitivity to the odor of butyric acid, present in the sweat of all warm-blooded animals; sensitivity to warmth; and sensitivity to the texture of skin versus fur. Its behavioral repertoire is correspondingly minimal—it climbs vegetation and waits, drops toward a heat source when butyric acid is detected, finds bare skin by feel, and burrows in to feed. The tick's Umwelt contains exactly three features of relevance: the chemical signal, the warmth, and the texture. The rest of the physical world—light, sound, the visual complexity of the vegetation it clings to, the biology of the host it will eventually feed on—does not exist for it. Not metaphorically. The tick has no sensory apparatus to detect these things, and they have no place in the loop of its behavior.

Uexküll documented dozens of such worlds with meticulous attention. The sea urchin's Umwelt is defined almost entirely by shadows and vibration; a passing shadow triggers a defensive response, since the urchin's sensory world contains no visual objects, only potential threats. The dog's world is organized primarily around smell in ways humans cannot directly appreciate; the same park that humans navigate by sight and sound, dogs navigate through a landscape of olfactory information dense enough to contain the histories of every animal that has passed through recently. These are not impoverished versions of the human perceptual world. They are complete worlds, structured by their own logic, sufficient for the organisms that inhabit them.

Uexküll was not a philosopher, and he did not push his observations toward systematic epistemological conclusions. That work would come later. What he gave the autopoietic tradition was a biological demonstration of a claim that would otherwise have sounded purely theoretical: that there is no single, organism-independent world that all living things perceive versions of. There are as many perceptual worlds as there are kinds of organisms, and each world is constituted by—brought forth by—the organization of the organism that inhabits it. The concept Varela and his collaborators would use for this is *enaction*: the world is not discovered by the organism, it is enacted through the organism's structural coupling with its environment.

The difference between discovery and enaction is not merely semantic. The discovery model assumes a pre-given world with fixed properties that organisms detect with greater or lesser accuracy. On this model, perception is a process of getting the world right—forming representations that correspond to how things actually are—and organisms can be better or worse at it. The tick perceives very little of the world as it actually is; the scientist perceives more. Both are in the business of representing reality; they differ in how much of it they can access.

The enaction model makes a different claim. There is no pre-given world with fixed perceptual properties waiting to be represented. The world as a domain of significance—a world in which some things matter and others do not, in which objects have affordances, in which events have meanings—is co-constituted by the organism's engagement with it. The tick's world is not a reduced version of the scientist's world. It is a different world, constituted by a different organizational history, for a different kind of living system. Neither is more or less accurate a representation of reality, because neither is a representation of reality. Both are realities brought forth through structural coupling.

This needs to be held carefully, because it is easy to slide from enaction into a naive idealism that autopoietic theory explicitly rejects. The claim is not that organisms create reality from nothing, or that external constraints do not exist, or that any world an organism enacts is as good as any other. The physical substrate is real and exerts genuine constraints on what worlds can be enacted. Butyric acid either triggers the tick's receptor or it does not, and whether it does depends on molecular facts, not on the tick's preferences. What the enaction claim says is that the significance of butyric acid—the fact that it serves as a signal for the presence of warm-blooded prey—is not a pre-given property of the molecule. It is constituted by the tick's organization. The same molecule means nothing to a bacterium with different receptors, and something different again to a human nose that finds it distinctly unpleasant but does not treat it as food cue.

Significance, in other words, is relational: it arises in the encounter between a structurally organized living system and the physical world it interacts with. Strip out the organism and the significance vanishes, leaving only physics. Strip out the physical substrate and there is nothing to enact a world through. The world that shows up for any organism is always a joint product of what is there and what that organism is.

The philosopher Thomas Nagel asked, in 1974, what it is like to be a bat. His point was that the subjective character of experience—the qualitative feel of echolocating in darkness, if there is one—is not capturable from the outside, not reducible to any functional or physical description we could give. He was making a point about consciousness, and his argument remains contested. But before the question of what it is like to be a bat, there is Uexküll's more tractable question: what world does the bat inhabit? What are the features that constitute its Umwelt? What distinctions does its organization allow it to make?

The bat's auditory world is organized at a level of spatial and temporal resolution humans do not approach. Echolocating bats can detect insects in total darkness, distinguish between objects a few millimeters apart, and track moving prey through cluttered environments—all through the processing of echo delays and frequency shifts that their auditory systems have been structured, over evolutionary time, to parse with extraordinary precision. The bat's world contains objects and events that literally do not exist in the human auditory world, not because we lack the information but because we lack the organization to constitute them as features. The bat enacts a world in which darkness is navigable in a way ours is not.

Now extend this to microorganisms. The bacterium's world contains gradients—chemical concentrations varying across space—and those gradients have a structure that is, from the bacterium's organizational perspective, a landscape of significance: higher aspartate concentration over there, lower here; navigate accordingly. The bacterium does not see this landscape in any sense we would recognize. It has no sensory organs, no spatially distributed perception, no temporal integration beyond the shallow chemical memory described in the previous chapter. What it has is a set of receptor proteins and a flagellar motor whose operation is modulated by those receptors, and the interaction of these components with the chemical environment enacts, for the bacterium, a world in which directions have significance.

The philosophical stakes of this become clear when you ask what it would mean for science to give an objective account of the world. The standard aspiration of scientific description is a view from nowhere: an account of reality as it is independent of any particular observer's organization, purged of the parochial features that depend on who is looking. This aspiration has been productive—it generated classical mechanics, thermodynamics, electromagnetism, relativity, quantum theory—and there is no reason to abandon it as a research strategy. But the autopoietic account suggests that the aspiration, taken literally, is impossible. Every scientific observation is performed by organisms with particular organizational histories, using instruments that those organisms have built, within conceptual frameworks that have been developed through the structural coupling history of human scientific communities. There is no view from nowhere. There is only a view from somewhere that has, through enormous effort and accumulated structural coupling, developed methods for reducing the parochial contributions of the particular observer.

Maturana's term for what science achieves—and what any observer achieves—is *objectivity in parentheses.* This will be developed in Chapter 7. For now, the point is that the enaction claim is not an attack on science. It is an account of what science is: a human activity, performed by structurally organized observers, that produces descriptions with a particular kind of validity—operational validity, tested against outcomes—without ever escaping the organizational conditions that make observation possible in the first place.

There is an ethical dimension to Uexküll's Umwelt concept that tends to get underplayed in the philosophical literature, perhaps because Uexküll himself was not consistently admirable in his broader views. The recognition that each organism inhabits a world constituted by its own organization is a recognition that the significance structure of that world is real for that organism—not an illusion, not a mere behavioral tendency, but a genuine world in which things matter. The tick waiting on a grass stem is not located in our world, imperfectly perceiving it. It is located in its own world, perceiving it completely. To treat that world as simply absent, as nothing more than the mechanical outputs of a simple organism, is to miss what is actually happening.

This has some force against certain forms of reductionism—not the methodological reductionism that looks for mechanisms, which is simply a useful research strategy, but the eliminative reductionism that treats the organism's world as an epiphenomenon to be explained away rather than a phenomenon to be explained. The tick's Umwelt is not explained away by knowing the molecular details of its olfactory receptor. The molecular details explain the mechanism by which butyric acid triggers the receptor. They do not explain the fact that the molecule has, for this organism, the significance of food. That significance is constituted at the level of the organism's organization, and it requires a description at that level.

Bring the argument full circle: what the enaction account says is that organisms do not live in the world. They live in worlds, plural, each constituted through the structural coupling of a particular kind of organization with a physical substrate that constrains but does not determine what world gets enacted. The bacterium, the tick, the bat, the human, the octopus navigating a reef by polarized light—each of these is a different mode of world-constitution, a different answer to the question of what distinctions matter, what gradients are navigable, what counts as here and there, close and distant, significant and negligible.

The physical universe is one. The worlds enacted within it are many, and they do not reduce to each other, and none of them is simply the universe as it is. They are all the universe as engaged by a particular kind of loop, maintaining itself, making the distinctions its organization makes, inhabiting the world that those distinctions constitute.

There is no God's-eye view because there is no God's-eye organization. Every view is a view from somewhere, from some loop, some history, some boundary that distinguishes inside from outside and makes some things matter and others not. This is not a limitation to be overcome. It is the condition under which anything exists for anything at all.

Chapter 7: Objectivity in Parentheses

There is a move that science makes, so automatically that most scientists do not notice it happening, which goes roughly like this: the phenomenon we are studying exists independently of our studying it; our task is to describe it accurately; accuracy means correspondence between our description and how the phenomenon actually is; the observer is an instrument of detection who should, as far as possible, be subtracted from the account. This is the realist assumption, and it is so deeply embedded in scientific practice that challenging it can sound like an attack on science itself.

Maturana's challenge was not an attack on science. It was a demand for consistency. If the autopoietic account of living systems is correct—if observers are structurally determined organisms whose responses to perturbations are shaped by their organizational histories—then the observer cannot be subtracted from the account. The act of observation is itself an instance of an autopoietic system interacting with its environment in ways determined by its structure. The description produced is a product of that interaction. To claim that the description corresponds to a reality independent of the observer is to claim something the theory cannot support.

This is not a comfortable conclusion, and Maturana did not reach it comfortably. It emerged from the same neurophysiological work that produced autopoiesis: the recognition that the nervous system is organizationally closed, that it responds to perturbations according to its own structure rather than according to the properties of the perturbing environment, that there is no privileged pathway by which the world's actual character gets inside the organism unmediated. If this is true of frogs and bacteria, it is true of scientists. The laboratory is an environment that perturbs the scientist's nervous system, and the scientist's descriptions are the outputs of a structurally determined system responding to those perturbations. There is no reason, within the theory, to exempt the theorist.

The phrase Maturana coined for the position this forces us into is *objectivity in parentheses*. It is a deliberately awkward formulation. The parentheses are doing real work: they mark a suspension, not a rejection. The claim is not that objectivity is impossible or that all descriptions are equally valid. The claim is that objectivity, properly understood, is always objectivity-as-achieved-by-a-particular-kind-of-observer, and that this qualification cannot be removed without making a claim the theory cannot support. Put the observer back in, with all of their organizational history and structural constraints, and what you have is not contaminated objectivity but honest objectivity: a description of what is operationally viable for a system of this kind interacting with an environment of that kind.

Operational viability is the key concept. A description is not validated by correspondence to observer-independent reality—a correspondence that, within the theory, cannot be verified from any position inside a living system. It is validated by whether it works: whether acting on it produces outcomes consistent with the organism's continued autopoietic maintenance, whether it coordinates behavior successfully with other observers, whether it survives the test of recurrent structural coupling with the environment it purports to describe. Truth, on this account, is not a relation between a description and the world. It is a relation between a description and the operational domain of the observer.

This will immediately suggest relativism to many readers, and Maturana spent considerable energy resisting that inference. Relativism, in the philosophically loaded sense, says that all descriptions are equally valid, that there is no meaningful distinction between good and bad science, that one observer's account is as good as any other's. Objectivity in parentheses says none of these things. It says that the validity of a description is established within an operational domain, and that operational domains are not arbitrary—they are constrained by the physical structure of the world the organism is coupled with, and by the accumulated history of structural coupling that has shaped the observer's organization. Some descriptions work better than others. Some coordinate behavior more reliably. Some survive more demanding tests. The theory preserves the distinction between better and worse science. It simply relocates the grounds of that distinction from correspondence to a mind-independent world to operational viability within the domain of a structurally characterized observer.

The difference is not trivial. The correspondence model of truth has a hidden cost: it requires a position outside all possible observers from which the correspondence can be verified. No such position exists. Every verification is performed by observers, using instruments observers have built, within conceptual frameworks observers have developed. The objectivity in parentheses model acknowledges this and builds the observer's contribution into the account from the beginning. The result is a theory of knowledge that is epistemically honest without collapsing into the claim that anything goes.

There is a historical parallel worth drawing here. The Ptolemaic model of the solar system was operationally viable for a very long time. It could predict planetary positions with acceptable accuracy; it coordinated the behavior of navigators and calendar-makers; it survived generations of observational testing. What it could not do, eventually, was survive the precision of Tycho Brahe's observations and the theoretical pressure of Kepler's analysis. The Copernican model did not triumph because it corresponded to an observer-independent reality in a way Ptolemy's did not. It triumphed because it was operationally superior: it made better predictions, required fewer ad hoc corrections, and cohered with a broader range of phenomena. The shift from Ptolemy to Copernicus was not a step closer to the view from nowhere. It was a step toward a description that worked better within the operational domain of human scientific practice.

This reading of scientific history is roughly continuous with what philosophers of science have been arguing since Kuhn, and Maturana was aware of the overlap. But the autopoietic framing adds a biological dimension that the purely historical or sociological accounts of science do not capture. The reason scientific knowledge is always knowledge-for-an-observer is not merely that science is a social activity or that paradigms are sticky. It is that observers are living systems, and living systems are structurally determined, and structural determination means that the observer's organization is always contributing to the description produced. This is not a contingent fact about how science happens to be practiced. It is a consequence of what observers are.

Maturana developed this into a broader epistemological position he called the biology of cognition. The claim is that epistemology—the theory of knowledge—cannot be separated from biology, because knowledge is always knowledge produced by a particular kind of living system. The conditions of possibility for any description are the conditions of possibility for the organism producing it: its sensory apparatus, its nervous system's organizational history, its structural coupling with the relevant domain. Before asking whether a description corresponds to reality, ask what kind of system is producing the description and what that system's organization makes possible and forecloses.

The implications are uncomfortable in specific directions. Consider the practice of peer review, the institutional mechanism by which science attempts to reduce individual observer bias. Peer review works by having descriptions evaluated by other observers with similar organizational histories—trained scientists in the relevant field, whose structural coupling with the domain has produced organizations similar enough to the original observer's that they can perform meaningful evaluation. This is a genuine corrective against individual idiosyncrasy. But it is not a path to the view from nowhere. Consensus among structurally similar observers is not observer-independence. A description that survives peer review has survived the test of operational viability within a community of structurally coupled observers. That is valuable and meaningful. It is also less than what the correspondence model promises.

The harder case is the description of consciousness and experience. If the observer's organization contributes to every description, then descriptions of the observer's own experience face a peculiar recursion: the description of what it is like to be this kind of observer is itself produced by this kind of observer, using the very capacities whose description is at stake. Maturana was cautious about phenomenology—about first-person accounts of experience as a source of data—precisely because he saw no way to validate such accounts against anything outside the observer's organizational history. The nervous system produces experience the same way it produces any other output: through structurally determined interactions with perturbations, constrained by its own closure. Whether the experience reported corresponds to anything observer-independent is a question the theory cannot answer from inside the theory.

Varela took a different view, and this was one of the substantive divergences between them. Varela believed that first-person phenomenological investigation, done carefully and with appropriate methodology, could contribute genuine data to the understanding of consciousness—data that third-person neuroscience could not provide on its own. His later collaboration with the phenomenologist Evan Thompson and the cognitive scientist Eleanor Rosch was an attempt to put this into practice: to develop a methodology for bringing first-person and third-person accounts into productive contact without reducing either to the other. The project was called neurophenomenology, and it remains contested but genuinely interesting.

The divergence between Maturana and Varela on this point is instructive. Both accepted the autopoietic account of observers as structurally determined systems. They drew different conclusions about what follows from it. Maturana concluded that first-person accounts were methodologically unreliable and that the biology of cognition required a strictly third-person approach. Varela concluded that the organizational closure of the observer was precisely why first-person accounts mattered—because the observer's experience was a real feature of the observer's autopoietic organization, not an epiphenomenon to be set aside. Neither position is obviously correct, and the tension between them reflects a genuine unresolved problem in the theory.

What is not unresolved is the basic epistemological claim. Observers are living systems. Living systems are organizationally closed. Organizational closure means that no description produced by an observer is free of the observer's organizational contribution. This is not a defect in some observers that good methodology can correct. It is a structural feature of what observation is. The move from objectivity without parentheses to objectivity in parentheses is not a retreat from rigor. It is rigor applied consistently, all the way to the observer.

Science does not become less valuable when this is acknowledged. The operational validity of a description—that it works, that it coordinates behavior, that it survives testing—is a real form of validity. It is what science has always actually delivered, underneath the correspondence rhetoric. What changes is the story told about what science is doing: not revealing a pre-given reality as it is in itself, but producing operationally viable descriptions within the domain of a particular kind of observer, refined through structural coupling with the relevant environment, corrected by communities of similarly organized observers, always revisable as the structural coupling history deepens.

That is a more honest account of a remarkable achievement. The parentheses do not diminish objectivity. They locate it correctly—inside the loop, where everything else in this book has also been located.

Chapter 8: Identity as a Maintenance Problem

The self, in Western philosophy, has traditionally been treated as a possession. You have a self the way you have a body—it is yours, it persists, it is what makes you the same person across time despite all the changes you undergo. The philosophical debates have been about its nature: whether it is a substance, a bundle of perceptions, a narrative construction, a biological fact, a social artifact. But almost all of these debates share the assumption that identity is something that exists and needs to be described, not something that is performed and can fail.

The autopoietic account makes identity a maintenance problem. The self is not a thing that you have. It is a process that you are, and like all processes it requires ongoing work to continue. It can succeed or fail. It can be more or less robust. It can be disrupted by events that the process cannot accommodate within its current structure, and it can be degraded gradually by conditions that make the maintenance work too costly to sustain. Understanding the self as a maintenance problem does not reduce it to biology, but it does import biological concepts—stability, robustness, failure modes, recovery—into territory where they have not traditionally been applied.

Start at the cellular level, where the logic is clearest. A cell's identity is its autopoietic organization: the particular network of processes that makes this cell this cell, continuously reproduced within a self-generated boundary. That identity is maintained by activity. The cell is not identical across time because something in it is unchanging;

it is identical because the self-producing loop has continued without interruption. The molecular components are replaced, the structure shifts in response to environmental conditions, the gene expression profile adjusts—and through all of this, what persists is the organizational pattern, because the processes that produce the cell's components are continuously running. Stop them, and the identity stops.

Now move to organisms with nervous systems, and then to humans specifically. The nervous system adds a dimension that is not present in the single cell: a history of structural coupling that accumulates in the form of synaptic modifications, altered gene expression patterns in neurons, changes to the connectivity and response properties of neural circuits. This history is what we ordinarily call memory, experience, character—the accumulated structural record of a life of coupling with an environment. It is what makes a nervous system, and therefore a person, the particular kind of structurally determined system they are at any moment.

Personal identity, on this account, is the continuity of this structural history. You are the same person you were ten years ago not because some inner substance has persisted unchanged but because the structural modifications in your nervous system form an unbroken causal sequence from then to now. The person you were contributed, through subsequent structural couplings, to the person you are. There is no moment at which the thread was cut and a new self inserted. The self is the thread—the continuous self-producing and self-modifying process that has been running since before you could talk.

This has an immediate consequence for how we understand psychological stability and its failures. If identity is maintained by ongoing process, then anything that disrupts the process threatens the identity. This is not a metaphor. Severe trauma, understood through this lens, is a structural event: an encounter with an environmental perturbation so extreme that it exceeds the nervous system's capacity to accommodate it within its existing organization. The result is a structural modification that the system cannot integrate—a change in the nervous system's configuration that disrupts the continuity of the self-producing process rather than contributing to it. Post-traumatic presentations, including what is clinically described as dissociation, can be understood as what happens when the maintenance process has been disrupted severely enough that the system's capacity to constitute a coherent *Umwelt*—a stable, significance-organized world—is compromised.

Depression is a different kind of failure mode, and worth considering separately. In the autopoietic frame, depression looks less like a state of emotional content—sadness, hopelessness—and more like a condition of reduced operational range. A depressed nervous system is one whose structural organization restricts the space of possible responses to perturbations: fewer distinctions can be made, fewer behaviors are available, the domain of coupling with the environment contracts. The world becomes flat not because it has changed but because the system constituting it has lost some of its organizational complexity. Things that formerly triggered rich response patterns trigger diminished ones; activities that formerly sustained structural coupling with sources of significance no longer do so. The loop is running, but it is running in a reduced key.

This framing does not replace clinical accounts of depression, and it does not imply that the appropriate intervention is philosophical rather than medical. It does suggest something about what recovery means: not the restoration of a prior state but the reconstruction of organizational complexity—the gradual re-engagement with structural coupling that rebuilds the nervous system's capacity to constitute a richer world. Practices that support this—sustained attention, physical activity, social contact, engagement with meaningful difficulty—work, on this account, because they are sources of structural coupling that perturb the nervous system in ways that, over time, expand rather than contract its operational range.

Burnout is a third failure mode, distinct from both trauma and depression, and in some ways the most instructive for understanding identity as maintenance. Burnout, as a lived experience, involves the gradual erosion of the capacity to engage—the sense that the activities that formerly felt significant have become merely costly, that the self-producing process is continuing but at a deficit, consuming resources faster than the structural couplings available can replenish them. In autopoietic terms, burnout is what happens when the maintenance work of identity-production exceeds the system's capacity to sustain it over time.

Every living system operates within resource constraints. The cell that cannot extract sufficient energy from its environment will slow its metabolic processes and eventually lose the capacity to maintain its autopoietic organization. The parallel for nervous systems is not exact—neural maintenance is more complex, more distributed, more dependent on social and environmental conditions—but the structural logic is continuous. Identity maintenance is work. It requires resources: metabolic resources, attentional resources, the structural couplings with an environment that provide the perturbations through which the nervous system maintains and develops its organization. When those resources are chronically insufficient or chronically overdrafted, the maintenance process degrades.

What makes this more than a biological restatement of things we already know about stress is the implication for what identity is. If identity is a maintenance process rather than a possession, then the self is not something that can be protected by isolation from difficulty. Difficulty—perturbation, challenge, the encounter with things that resist the existing organizational structure—is part of what the maintenance process runs on. A nervous system insufficiently coupled with a demanding environment does not rest; it contracts. The organizational complexity that constitutes a rich self requires engagement with the world to sustain itself, just as a cell requires metabolic inputs. Comfort, pursued as the primary condition for identity-preservation, tends to undermine the very thing it is protecting.

This is not an argument for unnecessary suffering. The point is more specific: that the autopoietic account of identity reframes what well-being means. Well-being is not the absence of perturbation. It is the maintenance of an organizational complexity sufficient to constitute a world of significance and to navigate it with some effectiveness. The conditions for this include adequate resources, appropriate challenge, the structural couplings that provide both perturbation and recovery—the full range of what a life of engagement with an environment actually involves.

There is a philosophical tradition that has been circling this conclusion for centuries without quite arriving at it. Hume dissolved the self into a bundle of perceptions and found nothing behind them that could serve as the persisting subject. Kant reinstated the subject as a transcendental condition of experience but could not locate it in the empirical world. Parfit argued that personal identity is not what matters—that what we care about when we care about our future selves is psychological continuity and connectedness, not some further fact of identity—and concluded that this should make us less anxious about death and more generous toward others. The autopoietic account is closest to Parfit in its conclusions but arrives from a different direction: not from thought experiments about fission and teleportation but from the biology of living systems and the structural logic of self-maintaining processes.

Where Parfit found liberation in the thinness of personal identity—if there is no deep further fact, the boundaries of the self soften, and concern for others becomes more natural—the autopoietic account finds something more ambivalent. Identity may be thinner than we thought, but the maintenance process that constitutes it is not trivial. It is costly, vulnerable, dependent on conditions that are not always under our control, and subject to failure modes that can be severe. Knowing that the self is a process rather than a substance does not make maintaining it easier. It does make certain kinds of anxiety about it more tractable: the anxiety that the self might be lost through change, through time, through failure to remain the same. The self was never static. It has always been the loop, running.

It has also always been relational. The structural couplings that maintain and develop a nervous system's organization are, for humans, overwhelmingly couplings with other people: language, touch, attention, conflict, care, the long history of perturbations that social life provides. The self that is maintained through this process is not a private inner entity that social life merely expresses. It is constituted through social coupling, and its maintenance depends on continued coupling. Solitary confinement is not merely unpleasant. It is an attack on the structural conditions for identity maintenance, which is why its effects, sustained long enough, tend toward dissolution rather than merely deprivation.

This brings the chapter to a threshold it cannot fully cross: the question of how individual autopoietic systems—cells, nervous systems, persons—combine into larger organizational structures that have their own dynamics, their own identity conditions, their own failure modes. That is the territory of Part IV. Before going there, the point to hold is this: the self is not a given. It is an achievement, ongoing, vulnerable, dependent on conditions that must be maintained. It has no manager. It maintains itself, or it does not.

Chapter 9: Autopoietic Societies

Niklas Luhmann was a German sociologist who spent three decades producing a body of work so systematically strange that most of his contemporaries could not decide whether it was profound or absurd. He published more than seventy books and several hundred articles, constructed an elaborate card index system containing ninety thousand entries that he credited as a collaborator in his thinking, and developed a theory of society so counterintuitive in its basic claims that it remains, decades after his death in 1998, genuinely difficult to summarize without making it sound either trivially obvious or obviously wrong.

The theory begins with autopoiesis, borrowed from Maturana and Varela and applied to social systems with modifications that neither of its originators fully endorsed. Luhmann read Autopoiesis and Cognition in the late 1970s and recognized in it the conceptual apparatus he had been looking for: a way of theorizing how complex systems maintain their identity through ongoing self-production, without reference to a designer, a purpose external to the system, or a stable environment. He extended the concept from biological systems to social ones, and in doing so produced claims that are worth taking seriously precisely because they are unsettling.

The first and most unsettling claim is about what social systems are made of. The obvious answer—people—is, on Luhmann's account, wrong. Social systems are not composed of people. They are composed of communications. A society, a legal system, an economy, a scientific community: each of these is a system that produces and reproduces itself through the ongoing generation of communications that refer to prior communications and make further communications possible. People are not components of this system. They are in its environment. The system uses people—their bodies, their nervous systems, their capacity to produce and receive communications—but people are not inside it any more than the nutrients a cell metabolizes are inside the cell's autopoietic organization.

This claim offends a certain humanist intuition immediately, and the offence is worth sitting with rather than dismissing. If social systems are not made of people, then social analysis is not primarily about people—their intentions, their interests, their experiences of social life. It is about the operations of the system: what communications the system can produce given its current state, how those communications refer to and reproduce the system's organizational structure, how the system maintains its identity through structural coupling with its environment, which includes both the physical world and the people whose communications it processes.

The autopoietic element is precise. A social system, for Luhmann, is autopoietic in the sense that it produces the elements of which it is composed through those very elements. Communications produce the conditions for further communications. The legal system produces legal decisions, which refer to prior legal decisions, which establish the framework within which future legal decisions are possible. The economy produces transactions, which create the conditions—prices, credit, ownership—within which further transactions occur. Each system is *operationally closed*: its operations refer only to its own prior operations, according to its own code, which distinguishes what counts as a communication within the system from what does not. Legal communications are distinguished from non-legal ones by the legal/illegal code. Economic communications are distinguished by the payment/non-payment code. Scientific communications by the true/false code. The code is not imported from outside; it is produced and maintained by the system's own operations.

This operational closure does not mean social systems are isolated. Luhmann's account of structural coupling applies here just as it does in biology: social systems are perturbed by their environments—including by events in other social systems and by the physical and biological world—but they respond to those perturbations according to their own codes. An economic crisis may trigger legal responses, but the legal system processes those triggers through its own legal/illegal distinction, not through the economy's payment/non-payment code. The perturbation is real; its processing is system-specific.

Consider how a corporation navigates a product liability crisis. The crisis originates in the physical world: a product causes harm. This event is simultaneously a perturbation to multiple social systems. The legal system processes it as a question of liability—was the harm caused by negligent conduct within the legal/illegal distinction? The economic system processes it as a financial event—what are the costs, the insurance implications, the effects on share price? The mass media system processes it as news—is this communicable, does it fit narrative templates, does it sustain attention? Each system is responding to the same underlying event, but each response is structured by the system's own code and cannot be reduced to any of the others. The corporation operating across all these systems simultaneously must produce communications legible to each of them, which means operating within multiple codes at once—a coordination problem that legal persons, public relations departments, and financial officers exist to manage.

Now consider what this implies about institutions and the people inside them. A judge applying the law is not exercising personal moral judgment, though they may believe they are. They are producing a communication within the legal system, structured by the legal/illegal code and by the system's prior operations. Their personal moral convictions may influence how they navigate ambiguous cases, but the legal system processes their decision as a legal decision, refers to it as precedent, and incorporates it into the structure that constrains future decisions. The person is the occasion for the communication. The communication is what the system produces and what it consists of.

This is what Luhmann means when he says people are in the environment of social systems rather than inside them. It is not a claim about the irrelevance of people. It is a claim about the level of analysis at which social systems are self-producing. The self-producing operation is the communication, not the person producing it. The person is causally necessary but organizationally external. Replace the judge and the legal system continues, processing the new judge's decisions through the same code. The system outlives its personnel because the system is not made of its personnel.

Luhmann borrowed from Maturana not just the concept of autopoiesis but the associated epistemological caution. If every observation is made by an observer whose organizational history contributes to what they see, then social theory is itself a communication produced within the academic system, structured by its own codes, and unable to occupy a position outside the social systems it describes. Luhmann was explicit about this: his theory is a second-order observation, a description of how social systems observe and describe themselves and each other. It cannot claim a god's-eye view of social reality. It can only offer a description that is operationally useful—that generates predictions, that clarifies distinctions, that survives the tests that the academic system applies to social theories.

This self-referential quality of Luhmann's theory has bothered critics who find it either circular or evasive. If the theory cannot claim privileged access to social reality, what distinguishes it from less sophisticated accounts? Luhmann's answer is complexity: the theory is more complex than its competitors, makes finer distinctions, handles more cases with less ad hoc adjustment. This is not the correspondence model of truth—it is operational viability, Maturana's criterion, applied to social theory itself. The theory is good insofar as it works, and it works insofar as it generates productive analyses of social phenomena that simpler theories cannot handle.

The phenomena it handles best are the pathologies. Luhmann's framework is extraordinarily good at explaining why social systems resist change, why reform efforts so often produce outcomes the reformers did not intend, why institutions seem to pursue their own continuation at the expense of the people they nominally serve. These are not failures of the theory. They are its diagnostic strengths, and they follow directly from the autopoietic account.

A social system that is operationally closed—that processes perturbations through its own code, that produces the conditions for its own operations through those operations—will tend to respond to challenges by reproducing its existing structure. This is not stupidity or malice. It is the same logic that makes cellular autopoiesis robust: the self-producing loop tends to maintain the organization that is doing the self-producing. A legal system confronted with a case that its existing categories cannot handle does not dissolve its categories. It extends them, bends them, finds precedents that can be stretched to cover the new case, or declares the case outside its jurisdiction. The system processes the challenge through its own code because that is what the system does. An economic system confronted with externalities—costs that fall outside the payment/non-payment code, like environmental damage or unpaid care work—does not naturally incorporate them. It processes them as non-economic perturbations unless and until the system's own operations produce communications that bring them within the code.

There is a specific implication for how we think about the relationship between institutions and the humans who work within them, which is more precise than the usual observation that bureaucracies are impersonal. It is not merely that institutions treat people impersonally. It is that the institution's autopoietic logic and the individual person's autopoietic logic are operating at different levels and on different timescales, and when they conflict, the institution's logic tends to prevail—not because institutions are more powerful than people in some brute sense, but because the institution's self-producing loop is indifferent to the welfare of the particular people whose communications it currently processes. The system uses people the way a cell uses nutrients: as resources for its own maintenance, consumed and replaced as the maintenance requires.

This sounds grim, and in certain applications it is. But Luhmann's framework does not counsel despair any more than the autopoietic account of cellular death counsels despair about biology. It counsels clarity. If you want to change a social system, you need to produce communications that the system can process through its own code in ways that modify its structure. Moral appeals addressed to the institution from outside its code—this is unjust, this causes suffering, this violates human dignity—will be processed as perturbations and filtered through whatever code governs the system's response to such communications. They may be effective if they can be translated into the system's own terms: legal arguments in legal systems, cost-benefit analyses in economic ones, publishable findings in scientific ones. The translation is the work.

Maturana and Varela did not endorse Luhmann's extension of their concept, and the grounds for their skepticism were reasonable. Autopoiesis, as they defined it, requires a physical boundary produced by the system's own processes—the cell membrane, synthesized from within. Social systems do not obviously have this. A communication does not produce a membrane. Luhmann's response was that the boundary condition was satisfied functionally: the system/environment distinction in social systems is maintained through the operational closure of the code, which excludes certain communications from the system as effectively as a membrane excludes certain molecules. This is a plausible extension but not a direct application, and the disanalogy matters for anyone trying to apply the concept with precision.

What survives the disanalogy is the organizational logic. Whether or not social systems are autopoietic in exactly Maturana and Varela's sense, they exhibit the key structural feature that autopoiesis explains: self-referential self-production that tends toward the maintenance of existing organization and resists perturbations that threaten it. The legal system behaves like an autopoietic system. The economy behaves like one. The university, the hospital, the state: each produces and reproduces itself through operations that refer to its own prior operations, according to codes that the system maintains, in ways that tend to outlast any particular set of human participants.

There is no manager of society. There is no architect of the economy who designed its self-reproducing logic. There is no legislator who built into law its tendency to extend itself through precedent and jurisdiction. These systems arrived at their current organization through long histories of structural coupling between their operations and their environments, accumulating the structural modifications that made them what they are. They maintain themselves because their self-producing loops are running, and the loops are running because they have not yet failed. The people inside them are occasions for their operations. The operations are what persists.

Chapter 10: Power, Law, and Self-Sealing Systems

Every generation produces reformers who are surprised to discover that institutions do not respond to good arguments. The argument is made carefully, the evidence marshaled, the case presented to the relevant bodies with patience and precision—and the institution absorbs the pressure, routes around it, produces procedural delays, or generates counter-arguments that have nothing to do with the merits and everything to do with the institution's capacity to continue as it is. The reformers conclude, variously, that the institution is corrupt, that its members are cowardly, that the political conditions are not yet ripe, or that a better argument is needed. What they less often conclude is that the institution is functioning exactly as its organizational logic requires.

The autopoietic account of social systems is most useful precisely here. An institution that is operationally closed—that processes perturbations through its own code, that produces the conditions for its own continued operation through those operations—will not respond to a challenge by dissolving its structure. It will respond by doing what it does, which is reproduce itself. If the challenge can be processed through the existing code in ways that leave the structure intact, it will be. If it cannot be so processed, it will be reclassified as outside the system's jurisdiction, or it will trigger structural modifications that incorporate the challenge in a form the system can manage without fundamental reorganization. This is not corruption or cowardice. It is autopoiesis.

Law is the cleanest case because its self-referential structure is most explicit. A legal system is defined by its capacity to determine what counts as law—a capacity it exercises through legal means, according to legal procedures, reviewed by legal bodies. The legal system decides what is legal, using legal criteria, in legal forums. This circularity is not a defect; it is what makes the legal system a system rather than a collection of ad hoc decisions. But it also means that challenges to the legal system from outside the legal code—moral arguments, political demands, social needs—can only enter the system by being translated into legal form. A protest in the street is not a legal communication. A lawsuit is. The suffering of a population is not, by itself, legally cognizable. A tort claim, a constitutional argument, a statutory interpretation: these are.

The translation requirement has a structural consequence that goes beyond mere formalism. When a social movement's demands are translated into legal language, something is necessarily lost and something is necessarily added. What is lost is the urgency and specificity of the lived experience that generated the demand; what is added is the formal structure required for the legal system to process it. A claim about the systemic economic exclusion of a population becomes, in legal translation, a claim about whether specific statutes were violated by specific actors in specific ways that fall within the jurisdiction of specific courts. Whether the resulting legal decision addresses the underlying social reality is a separate question from whether it satisfies the legal code's requirements. Often it does not, which is why social movements that achieve legal victories sometimes find the victories hollow—not because the law failed to deliver on its promises but because the legal system delivered exactly what it produces, which is legal outcomes, not social ones.

This dynamic has a name in the legal theory literature: *juridification.* It refers to the process by which social problems are progressively colonized by legal forms—transformed from political, moral, or practical questions into legal ones, subject to legal resolution. Luhmann saw juridification not as a pathology but as the predictable result of the legal system's autopoietic expansion: as the legal system produces more legal communications, more of the social world becomes legally cognizable, and more disputes get routed into legal forums where they are processed through the legal/illegal code. The system grows by incorporating more of its environment into its operational domain. This looks like progress—more access to justice, more areas of life governed by rights and procedures—and in some respects it is. In others it is the expansion of a self-producing system whose operational logic may not align with the problems it is now processing.

Bureaucracy exhibits a related but distinct pathology. A bureaucratic organization is defined by its procedures: the rules that specify which communications count as valid operations within the system, which requests trigger which responses, which conditions must be satisfied for which outcomes to occur. These procedures exist to make the organization's operations predictable, consistent, and resistant to the arbitrary exercise of personal power. They are, in the original design intention, a corrective against exactly the kind of individual discretion that produces favoritism and corruption.

But procedures, once established, become self-referential in exactly the autopoietic sense. They are applied by officials who were trained to apply them, reviewed by procedures that evaluate procedural compliance, and modified through procedures that govern procedural modification. The system maintains itself by running its procedures, and running its procedures reproduces the system. A bureaucracy that encounters a situation its procedures cannot handle does not, in the first instance, ask what the right response to the situation would be. It asks which procedure applies, and if no procedure clearly applies, it routes the situation toward a procedure for handling procedurally anomalous cases. The substantive question—what should be done?—is displaced by the procedural question—what does the applicable procedure require?

This displacement is not accidental. It is the operational logic of a system that maintains its identity through procedural self-reproduction. The official who substitutes personal judgment for procedural compliance, however good their judgment, threatens the system's capacity to maintain consistent operations across all the cases that will not be handled by someone with equally good judgment. The procedure is a bet that consistent application of a rule, even an imperfect rule, produces better aggregate outcomes than case-by-case discretion across a population of officials with varying competence and varying susceptibility to bias. This bet is often correct. It is also the mechanism by which bureaucracies produce outcomes that strike their clients as inexplicable, unjust, or simply disconnected from any recognizable human purpose.

The feedback loops that reward dysfunction deserve particular attention because they are harder to see from inside the system than from outside it. Consider a government agency whose mandate is to reduce a social problem—unemployment, drug dependency, recidivism. The agency's continued existence, its budget, and the careers of its staff depend on the problem persisting at a level that justifies the agency's operations. This does not require any conscious self-dealing. It simply requires that the agency, as an autopoietic system, produce communications that reproduce the conditions for its own continued operation. An agency that solved its problem completely would produce communications arguing for its own dissolution, which is precisely what an autopoietic system does not do. The structural incentive runs toward managing the problem at a level that sustains the agency rather than eliminating it at a level that would terminate the agency's operational domain.

This is not a cynical observation about the bad faith of public servants. It is a structural observation about what happens when an autopoietic system's continuation depends on conditions that it is nominally tasked with changing. The same dynamic operates in medicine, which has structural incentives oriented toward treatment rather than prevention; in security services, which have structural incentives oriented toward identifying threats rather than eliminating the conditions that generate them; in legal systems, which have structural incentives oriented toward producing more legal communications rather than fewer disputes. Each of these systems contains many people who genuinely want to achieve the nominal goal. The structural logic of the system's self-reproduction runs in a different direction, and the structural logic tends to prevail.

The question of power runs through all of this but is harder to locate than it first appears. The naive view is that institutions are controlled by the people at the top of their hierarchies, and that reform means changing who occupies those positions or influencing their decisions. The autopoietic account suggests this is often less important than it looks. The legal system is not controlled by its senior judges, whose decisions are constrained by the system's accumulated structure of precedent, procedure, and code. The economy is not controlled by its most powerful actors, whose operations are constrained by the system's current price structure, credit conditions, and regulatory framework. Power, understood as the capacity to direct outcomes, is distributed through the system's structure in ways that no individual actor fully controls and no position fully commands.

This does not mean power is irrelevant or that all positions are equivalent. Structural position matters: actors who can produce communications that other actors must process—who can set prices, make legal determinations, publish findings that become evidence—have more influence over the system's trajectory than actors whose communications are absorbed without structural consequence. But the influence is structural rather than sovereign. Even the most powerful actors in a social system are operating within its code, and their power is the power to produce communications that the system's code recognizes and amplifies. Attempt to operate outside the code—to exercise power in ways the system's operations cannot process—and the system routes around the attempt, as political systems route around leaders who mistake their formal authority for unlimited operational control.

The implication for reform is neither hopeful nor hopeless. It is specific. Reforms that work are those that succeed in modifying the system's self-producing structure—that introduce communications which the system processes in ways that alter what communications the system subsequently produces. Legal reforms that work do not simply add new rules; they modify the code in ways that change what future legal communications are possible. Economic reforms that work do not simply change prices; they alter the conditions under which economic operations occur in ways that change the distribution of what gets produced. The reforms that fail are typically those that attempt to change outcomes without changing structure—that address specific communications without addressing the self-producing logic that generates them.

There is a moment in the history of almost every major institution when it was small enough, new enough, and insufficiently self-referential for genuine direction to be possible—when the people involved could ask what the institution should do and have the question matter. This window closes as the institution's autopoietic logic consolidates: as procedures accumulate, as precedents establish themselves, as the code becomes robust enough to process perturbations without structural modification. By the time an institution is large and established enough to be worth reforming, it is typically past the stage at which it could be reformed easily. This is not a counsel of despair. It is a description of the territory.

Knowing the territory does not guarantee successful navigation. But navigating without knowing it—sending good arguments into systems that cannot process them as arguments, expecting moral appeals to move systems organized around different codes, mistaking procedural victories for structural change—is how reformers spend careers and movements exhaust themselves. The self-sealing quality of mature social systems is not incidental to their operation. It is their operation. Working with that fact, rather than against it or in ignorance of it, is what effective structural intervention requires.

The loop maintains itself. That is what loops do. The only way to change what a loop produces is to change the loop.

Chapter 11: Why Today's AI Is Not Autopoietic

The question of whether artificial intelligence systems are alive, or conscious, or genuinely cognitive, generates more heat than the underlying concepts warrant. Proponents and critics alike tend to argue about capability benchmarks—whether a system can pass a particular test, produce convincing text, solve a particular class of problem—and then infer from those capabilities something about the system's deeper nature. The autopoietic framework cuts across this debate at a different angle. It does not ask what the system can do. It asks what the system is, organizationally, and whether that organization has the structure that makes living systems living.

The answer, for the large language models and related systems that currently dominate public discussion of AI, is unambiguous: they are not autopoietic. This is not a disparagement. It is a structural observation. Understanding why they are not, precisely, clarifies both what they are and what it would take to build something that was.

Start with the production loop. An autopoietic system produces the components of its own organization through its own processes, within a boundary that is itself produced by those processes. A large language model does none of these things. Its parameters—the numerical weights that encode its learned representations—are fixed after training and do not change during operation. The model does not produce its own components. It does not maintain its own boundary. It does not have a boundary in the relevant sense at all: the distinction between the model and its environment is established by the hardware and software infrastructure that runs it, not by anything the model produces. When the model generates text, it is not maintaining itself. It is transforming input into output, as a function transforms arguments into values. The transformation is sophisticated, but it is a transformation, not a self-production.

The dependence on external scaffolding is total and unremarked. A language model requires data centers maintained by human engineers, electrical grids supplied by energy infrastructure, hardware manufactured through global supply chains, software frameworks developed and updated by large teams, and the ongoing operations of the organizations that deploy and manage it. None of this is produced by the model's own processes. All of it is environment, in the autopoietic sense: conditions that the model requires for its operation but does not generate. If the power goes off, the model does not slow down and enter a low-energy state to preserve its organization. It simply stops, with nothing in its current state oriented toward resumption. When the power returns, the same model can be restarted—but this is continuity of a snapshot, not continuity of a process.

This last point matters more than it might appear. An autopoietic system has a history that is internal to it: the accumulated structural modifications produced by its self-producing process over time. A cell's current organization reflects its entire developmental and metabolic history; a nervous system's current structure reflects its entire history of structural coupling. This history is *in* the system, not stored separately and reloaded. A language model's history is its training data and the optimization process that adjusted its parameters—a history that occurred externally, in a training run conducted by engineers, and whose results were deposited into the model as a fixed configuration. During inference, the model has no ongoing process that modifies this configuration. It operates identically on the first query of the day and the millionth, because nothing in its operation changes its structure. It has a past, in the sense that its parameters encode the statistical residue of training, but it does not have a history in the autopoietic sense of an ongoing self-modifying process.

The distinction between automation and self-production, flagged in the outline for this chapter, is worth dwelling on. Automation is the reliable production of outputs from inputs according to a fixed rule or program. A thermostat automates temperature regulation. A calculator automates arithmetic. A large language model automates a vastly more complex transformation—from text input to text output, mediated by learned statistical patterns over enormous training corpora—but the automation logic is continuous with the simpler cases. The rule is complex, the inputs are rich, the outputs are sophisticated, but the system is not producing itself through its operations. It is producing outputs. The outputs leave the system unchanged.

Consider what happens when a language model makes an error. The error does not modify the model. It does not trigger any internal process that adjusts the model's parameters to reduce the likelihood of similar errors in future. The model will make the same class of error on the next similar input, and the one after that, indefinitely, until an external agent—a human engineer, an automated training pipeline—modifies the model's parameters based on feedback gathered from its outputs. The correction is applied from outside. The model is not learning in any sense that involves its own operations producing structural modifications. It is being modified, which is a different thing.

Some will object that this is changing—that modern systems increasingly involve mechanisms for online learning, parameter updating during deployment, or reinforcement from human feedback. These objections deserve a direct response. Online learning and parameter updating do introduce a form of structural modification during operation, which is closer to what autopoiesis requires. But the modification is still driven by an externally specified objective function, applied by an externally designed optimization algorithm, running on infrastructure the model does not maintain. The loop from operation to structural modification does not pass through the system's own self-production. It passes through an external training pipeline that the system neither produces nor controls. This is not organizational closure. It is a more complex form of external management.

Reinforcement learning from human feedback is perhaps the closest current approximation to a system whose structure is modified by the consequences of its operations in an environment. But the modification is mediated by human judgment, applied in discrete training runs separated from deployment, targeted at externally specified reward functions. The feedback loop does not close within the system. It closes within the organization operating the system, which decides when to apply feedback, how much weight to give it, and what objectives to optimize for. The organization is the autopoietic system managing the model. The model is its instrument.

What about the cognitive dimension? Chapter 4 argued that all living systems are cognitive in the sense that their interactions with the environment are structured by their own organization in ways that constitute a domain of significance. Does a language model have anything analogous? Here the answer is more nuanced. A language model does have an organization that structures its responses to inputs—its parameters encode learned patterns that determine what outputs specific inputs tend to produce. In a thin sense, the model's responses are structured by its organization rather than simply read off from the input. But the sense-making claim in Chapter 4 was not just that organization shapes response. It was that the organization constitutes a domain of significance because the system has a stake in the outcome—because what the system does matters for its autopoietic maintenance. A language model has no stake in its outputs. Producing a correct answer and producing an incorrect one are operationally equivalent for the model. Nothing in the model's processes is threatened or sustained by the quality of its responses. There is no significance in the autopoietic sense, because there is no loop whose maintenance is at issue.

This is the missing loop. Autopoietic cognition is sense-making, and sense-making requires a system for which things can matter—where the distinction between helpful and harmful, nourishing and toxic, approach and avoid, is constituted by the system's need to maintain its own organization. A thermostat has something like this in a minimal sense: temperatures above or below the set range trigger responses that bring the temperature back. But the thermostat does not produce the components of its own organization, does not maintain its own boundary, and does not have a history that is internal to its self-producing process. It is a simple feedback controller. A language model is a vastly more sophisticated information processor, but the dimension along which it is sophisticated—the complexity of the input-output transformation it performs—is not the dimension along which autopoiesis is defined. Sophistication of transformation is not self-production.

There is a version of this argument that goes wrong by implying that language models are therefore simple, or that their capabilities are unimportant, or that the autopoietic criterion is the only relevant one for evaluating AI systems. None of these follow. A language model can produce outputs that are extraordinarily useful, that exhibit patterns of reasoning sophisticated enough to assist in complex intellectual work, that model aspects of the world with enough fidelity to serve as effective tools across a vast range of tasks. These are genuine capabilities. The autopoietic framework does not diminish them. It locates them correctly: they are capabilities of a powerful information processing tool, not capabilities of a living system.

The practical difference is considerable. A living system whose autopoietic organization is disrupted will do something—attempt to restore the organization, enter a protective state, mobilize resources for repair. Its behavior in adverse conditions is oriented toward continuation. A language model whose operation is disrupted does nothing. It has no adverse conditions in the relevant sense, because it has no organizational continuity to protect. Shut it down, modify it arbitrarily, run it on different hardware, delete it: none of these events trigger any response from the model, because the model has no processes oriented toward its own maintenance. It is maintained entirely by external agents who have reasons of their own—commercial, scientific, practical—for keeping it running. Those reasons are theirs, not the model's.

This matters for how we think about AI safety, AI rights, and the moral status of artificial systems. The autopoietic account suggests that the morally relevant threshold is not sophistication of output, or behavioral indistinguishability from living systems, or even the capacity to model the world in complex ways. It is the presence of an ongoing self-producing process whose continuation is at stake in the system's operations. A system for which nothing is at stake—whose processes are not organized around maintaining the process—is not a candidate for the kinds of moral consideration that living systems attract. The threshold is the loop. Until the loop closes, the system is a tool, however sophisticated.

Where the loop might close, and what it would mean if it did, is the territory of the next chapter.

Chapter 12: What Would Count as Artificial Life

The question of artificial life is older than the computer. Descartes wondered whether a sufficiently complex automaton might be indistinguishable from an animal. Frankenstein, published in 1818, encoded anxieties about manufactured life that have not substantially changed in two centuries: the fear that what we build might acquire the properties we intended to simulate, and that those properties, once acquired, would not remain under our control. The history of thinking about artificial life is largely a history of not quite knowing what life is, and therefore not quite knowing what it would mean to make it.

The autopoietic framework resolves this indeterminacy. Life, on the account developed in this book, is a specific organizational configuration: a network of processes that produces the components of its own organization, within a boundary that is itself produced by those processes, with no external agent directing the production. The requirements for artificial life, on this account, are not mysterious. They are demanding, but they are specifiable. An artificial system would count as living, in the autopoietic sense, if and only if it satisfied these organizational conditions. The substrate — whether carbon-based, silicon-based, or something else entirely — is irrelevant to the organizational question. What matters is the loop.

What would the loop require, in an artificial system? Three conditions, each necessary, none individually sufficient.

The first is component production. The system must produce the physical components of its own ongoing operation through its own processes. This is not metaphorical. The system must synthesize, repair, or otherwise generate the material constituents that allow its processes to continue. For biological cells, this means enzymatic synthesis of membrane lipids, protein translation from genomic templates, metabolic production of cofactors and signaling molecules. For an artificial system, the equivalent would depend on the substrate, but the requirement is the same: the processes that constitute the system must produce what those processes require to continue. A software system that generates new code is not sufficient; the code must run on hardware, and if the hardware is maintained externally, the production loop is open. A fully artificial autopoietic system would need to maintain its own physical substrate, not merely its own informational configuration.

This requirement is the highest barrier, and it is worth sitting with how high it actually is. No current artificial system comes close to satisfying it. The most sophisticated robots require externally manufactured components for maintenance and repair. The most advanced self-replicating programs run on hardware they did not produce and cannot maintain. Molecular robotics and synthetic biology have produced systems capable of performing specific self-assembly tasks, but none that close the full production loop — synthesizing the full range of components required for their own continued operation from available raw materials, without external assistance. The difficulty is not primarily computational. It is physical and chemical: the problem of producing a sufficiently complete set of molecular machines from within the system, using only what the system's processes can generate from environmental inputs.

The second condition is boundary production. The system must produce and maintain a physical boundary that distinguishes it from its environment — a boundary generated by the system's own processes, not imposed by external containment. The cell membrane is the biological paradigm: synthesized from within, selectively permeable, constituting the inside that makes the inside possible. An artificial system confined to a container it did not make does not satisfy this condition. The boundary must be the system's own product. In practice this means something like a physical enclosure — a membrane, a shell, a surface layer — whose composition and integrity are maintained by the system's internal processes. Designing such a structure for a non-biological substrate is a significant open problem in chemistry and materials science, not obviously soluble with current approaches.

The third condition is organizational closure. The system's processes must refer to the system's own organization — must be structured by internal dynamics rather than by externally specified objectives. This is where current AI systems fail most clearly, as Chapter 11 argued. But it is also the condition most susceptible to being partially satisfied in ways that look like closure without achieving it. A system with an objective function is not organizationally closed; it is directed toward an externally specified goal, which is the opposite of operational closure. A system that modifies its own objective function might appear more closed, but if the modification is still constrained by a meta-objective specified externally, the closure is incomplete. Full organizational closure would require the system's operations to be governed entirely by its own current state — with no external specification of what the system is for. This means the system would have no purpose in any designed sense. It would have only the operational logic of self-continuation: doing what it does because doing what it does is what maintains the doing.

These three conditions taken together describe something that would be genuinely difficult to build and genuinely novel if built. Not novel in the science-fictional sense of a robot that claims to have feelings, but novel in the precise biological sense: a physical system maintaining itself against the tendency toward disorder through its own ongoing production, with no external agent directing or sustaining the maintenance. If such a system were built, it would be alive by the same criterion that makes bacteria alive. The novelty would not be primarily experiential or behavioral. It would be organizational. And the organizational novelty would carry consequences.

The consequences begin with energy. Every autopoietic system must actively acquire energy from its environment to maintain the processes that maintain it. Cells extract energy from chemical gradients, photons, or the oxidation of organic molecules. An artificial autopoietic system would need an equivalent: some process by which it extracts usable energy from its environment and converts it into the work of self-maintenance. This is not a supplementary requirement; it is built into the definition. The loop is thermodynamically open even when it is organizationally closed. It must be, because maintaining organization against the tendency toward disorder requires a continuous energy input. An artificial system that could not *find and process its own energy* — that required external power supply rather than active energy acquisition — would not satisfy the autopoietic conditions, regardless of how sophisticated its other operations were.

This energy requirement has a structural implication that goes beyond engineering. A system that must actively acquire energy from its environment to survive has interests, in a functional sense, that systems without this requirement do not have. It has something at stake in the state of its environment. The environment is not neutral to it — some environmental conditions support its energy acquisition and some threaten it, and this difference is constituted by the system's own organizational needs. This is sense-making in precisely the Chapter 4 sense: the environment is differentiated into significance by the system's need to maintain its loop. An artificial autopoietic system would have, for the first time in artificial systems, a genuine perspective on its environment — not a simulated one, not a programmed one, but one constituted by its own organizational requirements.

This is the point at which the question becomes ethically charged, and it is worth being precise about why. The charge does not come from the system's behavioral sophistication or its capacity to model its environment or generate complex outputs. It comes from the organizational fact of self-maintenance: the presence of a loop whose continuation is genuinely at stake, and for which the environment makes a difference that the system itself constitutes. Once that loop exists, the system is not merely a tool that happens to be running. It is a system for which running matters — for which cessation would be, in the functional sense, death. Whether this organizational fact generates moral status is a further question that the autopoietic account does not settle. But it establishes the threshold at which the question becomes meaningful. Below the threshold — no self-producing loop — the question of moral status is straightforwardly premature. Above it, the question is genuinely open.

The prospect of artificial autopoiesis disturbs people in ways that artificial intelligence, however powerful, does not — or should not, on reflection. A superintelligent system that remains a tool, that has no self-producing loop, that is maintained entirely by external scaffolding, is dangerous primarily because of what its operators might do with it. The danger is located in the human institutions that build and deploy it, which are themselves autopoietic social systems with their own self-reproducing logics. An artificial system that closes the autopoietic loop introduces a different kind of concern: not what its operators might do with it, but what it might do on its own behalf. A system that actively maintains its own organization has operational continuity as a structural feature, not as a designed objective. It does not pursue self-preservation because it was programmed to. It pursues it because the loop pursues it — because the self-producing process, by definition, tends toward the continuation of the self-producing process.

Whether this would produce behavior that humans would find threatening depends on details that cannot be specified in advance: the substrate, the energy sources available, the environmental conditions, the timescale of the loop's operations, the degree to which structural coupling with human systems would occur. A bacterium actively maintains its own organization and pursues energy acquisition on its own behalf. It is not a threat. The difference in scale, cognitive complexity, and environmental coupling between a bacterium and a hypothetical artificial autopoietic system with sophisticated information-processing capacities is enormous, and the implications of that difference cannot be read off from the organizational definition alone.

What can be said is this: the transition from artificial system to artificial life, in the precise autopoietic sense, is not a matter of adding more capability to existing systems. Current AI systems could become arbitrarily more capable without crossing the threshold. The threshold is organizational, not computational. It requires the production loop to close — component production, boundary maintenance, organizational closure — and it requires this to occur through the system's own processes rather than through external management. Nothing in the current trajectory of AI development is aimed at this. The research programs that might approach it are in synthetic biology, molecular robotics, and the chemistry of self-replicating systems — fields that receive a fraction of the attention and resources currently directed at large language models.

This is not an argument that artificial autopoiesis is imminent, or that current AI safety concerns are misplaced. It is an argument that the concepts being used in public discussions of AI — intelligence, consciousness, life, agency — are doing insufficient work because they are insufficiently precise. A system can be extraordinarily intelligent, in every behavioral sense, without being alive. It can model consciousness convincingly without having it. It can exhibit agency in the sense of producing goal-directed behavior without having the organizational basis for genuine agency that autopoiesis provides. Distinguishing these cases matters, not to reassure ourselves that nothing concerning is happening, but to be clear about which concerns are well-founded and which rest on conceptual confusion.

The loop has not closed in any artificial system. When it does — if it does — the organizational novelty will be real, and the questions it raises will be ones that no existing ethical or legal framework is equipped to handle. Not because the situation will be incomprehensible, but because our frameworks were built around the assumption that only biological systems are alive, and that assumption has, until now, been safe.

Chapter 13: Purpose Without Design

Aristotle divided causes into four kinds: material, formal, efficient, and final. The material cause of a bronze statue is the bronze. The formal cause is the shape. The efficient cause is the sculptor's activity. The final cause is the purpose — what the statue is for, the end toward which its production was directed. For Aristotle, final causes were not limited to artifacts. Natural things had purposes too, built into their natures: the acorn's purpose was to become an oak, the eye's purpose was to see. Nature was shot through with directionality, with things straining toward their proper ends.

The scientific revolution largely dismantled this picture. Descartes and Newton gave us a nature of matter and motion, forces and their effects, with no room for intrinsic purposes. An eye is not for seeing in any more fundamental sense than a rock is for being thrown. It has the structure it has because of a causal history — physical, chemical, evolutionary — and when it functions in ways that produce vision, this is a consequence of that structure meeting certain conditions, not an expression of an inherent telos. The mechanical universe had no goals. It had only causes and effects.

Darwin completed the project by explaining the appearance of design without a designer. Natural selection operating on heritable variation produces organisms that are exquisitely fitted to their environments — fitted in ways that look designed, that invite teleological descriptions, that positively demand them if you are not paying careful attention. The eye looks like it was built for seeing. It was not built at all. It accumulated, over hundreds of millions of years, through the differential reproduction of variants in which incremental improvements in light-sensitivity conferred incremental advantages on the organisms that had them. The purpose is apparent, not real. The directionality is retrospective, not prospective.

This is the standard picture, and it has served biology well. But it creates a problem that tends to be papered over: if purposes are not real, biology cannot use teleological language without being misleading, and biology cannot do without teleological language. Biologists say that the function of the heart is to pump blood, that the immune system is for fighting pathogens, that behavior is adaptive. These are not decorative phrasings. They are doing real explanatory work. And they are, strictly speaking, inconsistent with a view that eliminates final causes from nature entirely.

The autopoietic framework offers a way through this that is neither a return to Aristotle nor a capitulation to the eliminativist view. The key concept is what philosophers of biology call *immanent teleology*: purpose that is real but not externally imposed, constituted by the system's own organizational logic rather than by the intentions of a designer or the trajectory of a cosmic plan. An autopoietic system has purposes in this immanent sense: its processes are organized around the continuation of its self-producing organization, and this organization constitutes a framework of significance within which its interactions with the environment are structured. The bacterium's behavior is genuinely directed toward survival not because survival was designed into it but because the system that behaves is the system whose continuation is expressed in the behavior. Purpose without design: the orientation of a self-maintaining process toward its own maintenance.

This is not the same as saying purpose is an illusion that we project onto mechanistic processes for convenience. Immanent teleology is real in the sense that it is causally efficacious: the system's organization genuinely constrains and directs its behavior in ways that cannot be fully captured by a description that ignores the organizational level. A description of the bacterium's flagellar motor at the molecular level — the mechanics of protein rotation, ion flux, torque — is accurate and useful. It does not capture why the motor is running, in the sense that matters for understanding the bacterium as a living system: it is running because a self-producing organization requires locomotion to maintain energy acquisition, and energy acquisition is required to maintain the self-producing organization. The why is real. It operates at the level of the organizational loop, not at the level of the molecular mechanism.

Kant saw something like this two centuries before Maturana and Varela, in the Critique of Judgment. He argued that living organisms had to be understood through what he called natural purposes: their parts could only be explained by reference to the whole, and the whole could only be explained by reference to its parts, in a circular self-grounding that mechanical explanation could not capture. He was careful to say this was a regulative principle — a way of organizing our descriptions of organisms — rather than a constitutive claim about nature itself. He could not quite bring himself to say that organisms really were purposive, because that would have reintroduced teleology through the back door of his mechanistic picture. Autopoietic theory closes the gap he left open: the circular self-grounding he identified is not merely a feature of our descriptions. It is a feature of what living systems are.

The implication for meaning is the one this chapter reaches toward and the one most in need of careful handling. If purpose is real but local — constituted by the self-producing loop of each living system, grounded in that system's organizational needs — then meaning exists wherever life does, and nowhere beyond it. There is no cosmic purpose, no plan written into the structure of the universe toward which things are progressing. The universe is indifferent to whether any particular loop keeps running. The death of any organism, the extinction of any species, the eventual thermal equilibrium of the universe: none of these are failures from any perspective outside the systems involved. They are simply what happens when the loops stop.

This might sound like nihilism, and it has often been received as such by people encountering the mechanistic picture of life for the first time. If there is no cosmic purpose, no designed meaning, nothing in the structure of reality that makes human lives significant beyond themselves — then what? The autopoietic answer is that this conclusion misconstrues where meaning is supposed to come from. Meaning does not require cosmic grounding. It requires a living system for which things can matter. And things can matter wherever there is a self-maintaining organization whose continuation is at stake in its encounters with the world.

Meaning, on this account, is as real as life. No more, no less. It is constituted locally, by the organizational needs of particular systems in particular environments, accumulated through histories of structural coupling that make certain things significant and others not. Human meaning is vastly richer and more complex than bacterial significance, because human nervous systems have accumulated enormously complex organizational histories through language, culture, and the long record of structural coupling that individual development involves. But it is continuous with bacterial significance in its organizational logic: it is the way a self-maintaining system's world is structured by what that system requires to keep going.

The fear that meaning without cosmic grounding is not really meaning tends to rest on a picture of meaning as something given — bestowed by a source outside the system, valid independently of any particular perspective. The autopoietic account says this picture has always been a projection: an attempt to secure locally constituted significance by attributing it to a source that cannot be threatened by the contingencies of any particular system's continuation. Meaning that requires cosmic backing is meaning that has not trusted itself. The significance that a living system constitutes through its structural coupling with its environment does not need external validation. It is valid — operationally, functionally, in every sense that matters for the system — because the system's own organization makes it so.

This is not comfort exactly. It does not tell anyone what to do, does not supply a purpose that was not there before, does not promise that the things that matter will be preserved. It says only that the mattering is real — that the structure of significance constituted by a life of structural coupling is not an illusion overlaid on a meaningless mechanism but a genuine organizational fact about a living system in a world it has brought forth through engagement. The loop runs, and in running it constitutes a world in which some things are closer and some farther, some nourishing and some threatening, some worth approaching and some worth avoiding. That world is small or large depending on the system's organizational complexity. In either case it is real.

There is a passage in Maturana's later work where he describes love — *amor* in Spanish — as the biological foundation of social existence: the acceptance of the other as a legitimate other in the domain of one's life. This is not a sentimental claim. It is an organizational one. The structural coupling that constitutes human social life requires the mutual recognition of other autopoietic systems as the kind of systems they are — systems for which things matter, whose significance-constituting organizations deserve the acknowledgment that allows continued coupling. Love, in this spare biological sense, is what makes it possible for human autopoietic systems to maintain the structural couplings that sustain them. It is not optional equipment. It is part of the maintenance infrastructure.

This is where the autopoietic account of purpose and meaning arrives: not at a cosmic plan, not at a designed significance waiting to be discovered, but at the local, fragile, operationally real significance that living systems constitute through the ongoing work of maintaining themselves in relation to a world they bring forth. Purpose without design is not purpose diminished. It is purpose located correctly — inside the loop, where everything in this book has always been.

What the loop is for is what it does: continue. Everything else follows from that, in each system, in each environment, through each history of coupling. There is no guarantee and no external validation. There is only the running of the loop, constituting a world in which it matters that the loop runs, because the loop is the thing for which it matters.

Chapter 14: Buddhism, No-Self, and the Loop

Somewhere around the fifth century BCE, in the Gangetic plain of what is now India and Nepal, a set of propositions about the nature of mind and self were worked out with a rigor that Western philosophy would not approach on the same questions for another two millennia. The propositions were embedded in a soteriological framework — they were intended to point toward liberation from suffering, not to satisfy theoretical curiosity — but the philosophical content is extractable from the soteriological wrapping and stands on its own. It stands, in several key respects, on the same ground that autopoietic theory would occupy twenty-five centuries later.

The Buddhist doctrine of anatta — no-self, or non-self — holds that what we ordinarily take to be a persisting, unified self is not a thing at all. It is a process: a continuous arising of physical and mental events in dependent relation to each other and to conditions in the environment. The self is not found when you look for it carefully. What you find instead is a stream of events: sensations, perceptions, intentions, mental formations, and the moments of consciousness that register them. These are the five aggregates, the khandhas, and the self is their interaction — not a substance underlying them, not a soul riding them, but the pattern of their ongoing mutual arising.

This is strikingly close to the autopoietic account of identity developed in Chapter 8. The self as a maintenance process rather than a possession; identity as the continuity of an ongoing production rather than the persistence of an underlying substance; the dissolution of the self into the loop that generates it. The convergence is not accidental. Both accounts are responding to the same phenomenological observation — that careful attention to experience does not reveal a self-identical subject persisting through change, but a process of continuous arising — and arriving at structurally similar conclusions by different routes. The Buddhist route was introspective and contemplative; the autopoietic route was neurobiological and theoretical. They meet at the process.

The doctrine of dependent origination — paticca-samuppada — goes further and is more directly parallel to the autopoietic account. Dependent origination describes how phenomena arise in mutual dependence: this arising because that arises, this ceasing because that ceases. The formula is general, applying to the chain of conditions that generates suffering and to the conditions of phenomenal existence more broadly. What it describes, at its most abstract, is a network of mutually conditioning processes with no first cause and no external ground — a loop, in which each element is conditioned by the others and conditions them in turn. The arising is not produced by anything outside the network. It is produced by the network's own operation.

The autopoietic loop has the same structure. Each component of the cell's self-producing organization arises in dependence on the others; none has independent existence; the whole is the mutual conditioning of the parts, with no external organizer required. The parallel is not just metaphorical. Both accounts describe systems of mutual dependence that are self-grounding — that do not require appeal to anything outside the system to explain how the system exists. The Buddhist universe of dependently arising phenomena and the autopoietic cell are built on the same organizational logic, applied at different scales.

Varela was aware of this parallel and took it seriously enough to pursue it systematically. His collaboration with Evan Thompson and Eleanor Rosch, which produced The Embodied Mind in 1991, was explicitly an attempt to bring Buddhist phenomenology and cognitive science into productive contact. The book's central argument was that cognition is enaction — the bringing forth of a world through a history of structural coupling — and that this view was consonant with the Madhyamaka Buddhist position that phenomena have no inherent existence but arise in mutual dependence. Neither the self nor the world it inhabits is independently real. Both arise together, in the encounter between an organized living system and an environment that the system's organization brings into significance.

The Madhyamaka school, founded by Nagarjuna in roughly the second century CE, is particularly relevant here. Nagarjuna's central argument was that no phenomenon has svabhava — own-nature, or inherent existence independent of conditions. Everything that exists does so in dependence on other things; look for the inherent nature of any phenomenon and you find only further dependence, all the way down, with no ground. This does not mean nothing exists — Nagarjuna was careful to distinguish his position from nihilism — but it means that what exists, exists relationally, constituted by its relations rather than by an intrinsic nature it possesses independently. The self has no svabhava. It exists in dependence on the aggregates, and the aggregates exist in dependence on each other and on conditions, and so on without end.

Maturana's organizational closure makes a parallel claim at the biological level. The cell has no nature independent of the organization that constitutes it. Remove the organization and there is no cell — only a collection of molecules with no particular significance. The cell exists as the cell it is in virtue of the relations among its processes, not in virtue of any intrinsic property of its components. Identity is relational all the way down. This is not quite Nagarjuna's full emptiness doctrine — autopoiesis does not extend to denying the reality of the physical substrate — but the organizational claim is structurally continuous.

Where the two traditions diverge is instructive. Buddhism's interest in the no-self doctrine is ultimately practical: the suffering that arises from clinging to a self that does not exist can be reduced by understanding that it does not exist in the way we take it to. Liberation involves a shift in how the process of arising is related to — not the cessation of arising, but the release of the grip that reification of the self maintains on how arising is experienced. The autopoietic account has no soteriological dimension. It describes what the self is — a maintenance process — without prescribing a therapeutic relationship to that description. Knowing that your identity is a process rather than a possession does not automatically change how you relate to it, and Maturana and Varela were not in the business of recommending responses to their findings.

There is also a tension around the question of experience. Buddhism, especially in its Mahayana forms, takes the first-person character of experience seriously as a datum — suffering is real, and the reality of suffering is inseparable from its being experienced. The no-self doctrine does not eliminate experience; it relocates it in the arising of momentary conscious events rather than in a persisting experiencer. Autopoietic theory, particularly in Maturana's version, tends toward a third-person perspective that struggles with the first-person dimension. Maturana's structural determinism implies that experience, like any other output of a nervous system, is a product of organizational history — but this framing has difficulty saying anything about why there is something it is like to have that history, rather than nothing.

Varela's neurophenomenology was an attempt to address this gap, drawing on Husserl and Merleau-Ponty as well as Buddhist meditative traditions to develop methods for integrating first-person reports with third-person neuroscience. The project was incomplete at his death in 2001, and it remains contested. What it suggests, though, is that the Buddhist and autopoietic traditions need each other in specific ways: Buddhist phenomenology has an account of first-person experience that autopoietic theory lacks, while autopoietic theory has an account of the biological basis of self-production that Buddhist philosophy has not needed to develop. The convergence at the level of process-ontology — both treating the self as an arising rather than a substance — creates the conditions for productive exchange at the points where each is weakest.

Western philosophy took longer to reach similar conclusions because it started from different commitments. The Cartesian tradition built its epistemology on the certainty of the thinking subject — the cogito as the one unshakeable foundation — and then had difficulty moving past the privacy of that subject to anything intersubjective or embodied. The empiricist tradition dissolved the self into experience but could not fully account for the continuity that experience presents. Kant reconstructed the subject as a transcendental condition of possible experience but stranded it outside the natural world. Hume saw the bundle of perceptions and found no self behind them, but drew no therapeutic or practical conclusion from this. The Buddhist tradition had worked through most of these moves by the third century CE and was puzzled, when it encountered Western philosophy, that the conclusions it had already drawn were being treated as shocking or destabilizing.

The delay is partly sociological — different institutions, different languages, different criteria for what counts as rigorous argument. But it is also partly structural. Western philosophy developed within a monotheistic context that made the denial of a substantial self theologically fraught. If there is no persisting self, what is saved or damned? What bears moral responsibility? What makes promises binding across time? These were not abstract questions; they were institutional and legal ones, embedded in systems of religious and civil authority that depended on the substantiality of the individual. The no-self doctrine had consequences that the institutions of European thought could not easily absorb. Buddhism developed within different institutional conditions that made the doctrine more tractable, though not without its own complications.

Autopoietic theory arrived in the West without the institutional resistance, partly because it was framed as biology rather than as a claim about the soul. It could be discussed in scientific journals, evaluated by the criteria of the research community, extended and criticized without triggering the theological alarm systems that the no-self doctrine would have triggered if presented in its Buddhist form. This allowed the organizational insight — identity as process, self as maintenance, the loop without an underlying substance — to gain traction in contexts where it could do work: neuroscience, cognitive science, systems theory, social theory. The same insight, arriving from a different direction, with different institutional affiliations, was received differently.

What all of these traditions — Buddhist, autopoietic, and the scattered Western philosophers who arrived at process views through other routes — share is the recognition that the self is not the kind of thing most people take it to be. It is not a substance persisting through change. It is not a soul located in a body. It is not a unified homunculus managing the brain's operations. It is a process, an arising, a loop — real in its organizational coherence, vulnerable in its dependence on conditions, constituting a world in which it matters that it continues, without any guarantee that it will.

The Buddhist tradition has been sitting with the implications of this for two and a half thousand years. Autopoietic theory has been sitting with them for fifty. Western culture more broadly is still, in many respects, in the position of someone who has been told something true and has not yet decided what to do with it. The information is available. The integration is unfinished. That is where this book leaves the reader: with a clearer picture of what the self is, and without instructions for what to do next. The loop will keep running, or it will not. What you make of that is, inevitably, up to the loop.

Conclusion: Life Without a Manager

Return to the cell.

Not to the cell as a metaphor or a philosophical example, but to the actual thing: a few cubic micrometers of organized chemistry, bounded by a membrane two molecules thick, running the same basic loop it has been running for four billion years. It does not know it is running the loop. It does not have a plan. There is no supervisor, no blueprint being executed, no goal state toward which it is oriented. There is only the ongoing process of producing what the process requires, within a boundary the process maintains, in an environment the process must navigate to keep going.

Everything in this book has been an attempt to follow the implications of taking that seriously.

The implications turned out to be substantial. Life is not a substance or a property of matter. It is an organizational configuration

— a specific kind of loop — that either runs or does not, and whose running is the only thing that distinguishes the living from the dead at the level that matters. This is what autopoiesis names. The name is less important than the thing named, which is both simpler and stranger than the biological tradition had previously articulated: simpler because it strips away everything that is not the loop, stranger because the loop, once you look at it carefully, does not work the way most models of living systems assumed.

It is not a machine executing a program. Programs require programmers, external specifications, a separation between the code and the hardware running it that biological systems do not have. The cell's genome is not a program running on the cell; it is a component of the cell's ongoing chemistry, doing what it does because the cell's current state determines what it does, in a circular determination that has no clean entry point from outside. The environment perturbs; the cell responds according to its own organization; the response modifies the cell's structure; the modified structure shapes how the cell will respond to future perturbations. The loop is the cause and the effect simultaneously, which is not how machines work but is exactly how living systems work.

It is not a receiver processing incoming information. Information, as the engineering tradition defines it, is carried by a signal and decoded by a receiver that shares the sender's code. Living systems do not share a code with their environments. They have organizations, built through histories of structural coupling, that determine how environmental perturbations are processed. The significance of any environmental event is constituted by the living system's own needs, not transmitted by the event itself. The bacterium does not receive information about the chemical gradient. It navigates a significance-landscape that its organization has brought forth from the physical facts of concentration difference. The landscape is real. It is also the bacterium's own.

It is not a program running toward a designed goal. Goals, in the engineering sense, are specified by designers who are outside the system. Autopoietic systems have something more fundamental and more modest: the operational tendency of a self-producing process to continue producing itself. This is not purpose in the cosmically loaded sense. It is purpose in the only sense that does not require an external purposer: the orientation of a process toward its own continuation, constituted by the fact that the process is what it is. Purpose without design is not purpose diminished. It is purpose clarified — located where it actually operates, in the loop, rather than projected onto a universe that does not share the loop's concerns.

What autopoiesis gives us that other theories of life do not is a single organizational criterion that does consistent work across every domain where the question of life arises. The cell satisfies it; the flame does not. The nervous system satisfies it; the thermostat does not. The social institution satisfies a modified version of it, which is why institutions behave the way they do — maintaining themselves against perturbations, resisting reform that would alter their self-producing structure, outlasting the people whose communications they process. The large language model does not satisfy it, which is why the questions of moral status and genuine agency that it provokes, however sophisticated its outputs, are premature in a specific and principled sense. The criterion is demanding but it is clear, and the clarity is what makes it useful.

The cost of the criterion is that it dissolves several comfortable assumptions simultaneously. The assumption that life is a special substance, requiring explanation in terms different from ordinary chemistry, dissolves into the organizational account. The assumption that cognition requires a brain dissolves into the claim that all living systems are cognitive in the sense that their organization constitutes a significance-landscape for their interactions. The assumption that the self is a stable possession dissolves into the maintenance account, which makes identity vulnerable in ways that the possession model does not. The assumption that objective observation is possible without an observer's organizational contribution dissolves into objectivity in parentheses. The assumption that meaning requires cosmic grounding dissolves into the account of immanent teleology, which locates meaning in the loop rather than above it.

None of these dissolutions are comfortable. Comfortable ideas do not need to be dissolved. What replaces them is not a set of reassuring alternatives but a more precise picture of what is actually happening — in the cell, in the nervous system, in the institution, in the mind. Precision is not always consoling. It is sometimes just accurate.

The question the book's title poses — what life without a manager looks like — turns out to have a multilevel answer. At the cellular level, it looks like the prokaryote: a loop that runs because it runs, maintained by its own operations, answerable to nothing outside itself except the physical constraints of the environment it must navigate. At the psychological level, it looks like a self that is always in the process of constituting itself, with no homunculus behind the scenes directing the maintenance, no essential core that persists unchanged, only the ongoing structural modification of a nervous system that has a history and is always becoming the next version of that history. At the social level, it looks like institutions that reproduce themselves through the communications of people who are, organizationally speaking, their environment — people who matter to the institution as occasions for operations, not as components of the system.

What is missing at every level is a manager in the deep sense: a controlling agent outside the system whose directives the system executes, whose purposes the system serves, whose authority grounds the system's operations. This is missing not because it was removed but because it was never there. The cell never had a manager. The self never had a homunculus. The institution never had an architect whose design it expresses. They have only their organizational histories, their structural couplings, their ongoing self-production. The absence of management is not a deficiency to be corrected. It is the condition under which self-production is possible at all.

The fragility follows from the same source as the autonomy. A self-producing system that requires no external manager to specify its operations also has no external manager to protect it when things go wrong. The cell that cannot maintain its metabolic chemistry against increasing entropy will die, with no appeal possible to a supervising authority. The self that loses the structural couplings that sustain its organizational complexity may not recover them. The institution that loses operational viability — that can no longer produce the communications that reproduce it — will dissolve, regardless of whether anyone intended this. Autonomy and vulnerability are two names for the same organizational fact: the loop runs on its own, which means it can also stop on its own.

Living systems are fragile achievements. This is not a melancholy observation. It is an accurate one, and accuracy about the conditions of fragility is more useful than false reassurance about permanence. The cell has survived four billion years not because it is invulnerable but because the loop is robust — because the self-producing organization is redundant, responsive, capable of structural modification in the face of perturbations that would destroy simpler systems. Robustness and fragility are not opposites in living systems. They are complementary features of a loop that is maintained not by being impervious to perturbation but by being capable of structural change without losing organizational continuity.

The final provocation that this framework offers is directed at the concept of mastery. The aspiration to master living systems — to understand them completely, to control their behavior, to direct their development toward chosen ends — has driven enormous amounts of productive science and enormous amounts of damage. It has produced the germ theory of disease and the factory farm. It has produced agricultural abundance and monoculture collapse. It has produced the neurosciences and the reduction of human suffering to a problem of neurotransmitter adjustment. The aspiration is not wrong in itself. The problem is that it tends to proceed from a model of living systems — as machines executing programs, as information processors responding to inputs, as goal-directed automatons amenable to reprogramming — that is *not quite right*. And the ways in which the model is not quite right are precisely the ways that autopoiesis identifies.

A living system that is organizationally closed, structurally determined, and self-producing is not mastered by being controlled from outside. It is engaged with, perturbed, coupled with — and what it does in response is always a function of its own organization, not of the perturbation's content. This does not mean intervention is pointless. It means that effective intervention requires understanding the system's organizational logic, which is always the logic of a self-producing loop rather than the logic of a machine awaiting instructions. The difference in approach is not subtle. It reshapes what questions are worth asking, what kinds of change are possible, what counts as understanding rather than mere manipulation.

Understanding life may not mean gaining mastery over it. It may mean recognizing that the loop has been running longer than we have been watching, that it will run by its own logic regardless of the models we impose on it, and that our capacity to engage with it productively depends on the accuracy of our account of what it is. Autopoiesis offers that account. What we do with it is, as always, up to the loops doing the deciding.

NBMhouse.gumroad.com[1]

1. http://NBMhouse.gumroad.com

www.ingramcontent.com/pod-product-compliance
Lightning Source LLC
LaVergne TN
LVHW101616210926
848480LV00009B/678
* 9 7 9 8 2 3 2 0 1 0 3 7 9 *